DISCOVERY ENGINEERING

IN PHYSICAL SCIENCE

Case Studies for Grades 6–12

DISCOVERY ENGINEERING
IN PHYSICAL SCIENCE
Case Studies for Grades 6–12

M. GAIL JONES • ELYSA CORIN • MEGAN ENNES

EMILY CAYTON • GINA CHILDERS

National Science Teachers Association

Arlington, Virginia

National Science Teachers Association

Claire Reinburg, Director
Rachel Ledbetter, Managing Editor
Andrea Silen, Associate Editor
Jennifer Thompson, Associate Editor
Donna Yudkin, Book Acquisitions Manager

ART AND DESIGN
Will Thomas Jr., Director
Joe Butera, Senior Graphic Designer, cover and
    interior design

PRINTING AND PRODUCTION
Catherine Lorrain, Director

NATIONAL SCIENCE TEACHERS ASSOCIATION
David L. Evans, Executive Director

1840 Wilson Blvd., Arlington, VA 22201
www.nsta.org/store
For customer service inquiries, please call 800-277-5300.

Library of Congress Cataloging-in-Publication Data
Names: Jones, M. Gail, 1955- author.
Title: Discovery engineering in physical science : case studies for grades 6-12 / M. Gail Jones [and four others].
Description: Arlington, VA : National Science Teachers Association, [2019] | Includes bibliographical references.
Identifiers: LCCN 2018046696 (print) | LCCN 2018049966 (ebook) | ISBN 9781681406183 (e-book) | ISBN 9781681406176 (print)
Subjects: LCSH: Discoveries in science--Study and teaching (Elementary)--Activity programs. | Discoveries in science--Study and teaching (Middle school)--Activity programs. | Discoveries in science--Study and teaching (Secondary)--Activity programs. | Engineering--Study and teaching (Elementary)--Activity programs. | Engineering--Study and teaching (Middle school)--Activity programs. | Engineering--Study and teaching (Secondary)--Activity programs. | Science--Study and teaching (Elementary)--Activity programs | Science--Study and teaching (Middle school)--Activity programs. | Science--Study and teaching (Secondary)--Activity programs.
Classification: LCC Q180.55.D57 (ebook) | LCC Q180.55.D57 D57 2019 (print) | DDC 507.1/2--dc23
LC record available at *https://lccn.loc.gov/2018046696*

*This book is dedicated to all the youth who remind us
that the smallest things can be the most important.*

Contents

About the Authors...ix

Acknowledgments xi

Introduction...1

1   Don't Get Bent Out of Shape
Memory Wire   11

2   Harnessing the Color of Nature
Structural Color and Iridescence   25

3   From Sludge to Nobility
Artificial Dye   43

4   By the Teeth of Your Skin
Shark Skin and Bacteria   57

5   A Sticky Situation
Gecko Feet Adhesives   73

6   It's Hip to Be Square
Seahorse Tails   89

7   Corn Flakes
Waste Not, Want Not   105

8   Get Fired Up With Friction Lights
Matches   125

9   That's a Wrap
Plastic   139

10   A Sticky Discovery
The Invention of Post-It Notes   153

11   Putty in Your Hands
The Unintended Discovery of Silly Putty   167

12   Sweet and Sour
The Rise and Fall ... and Rise ... of Saccharin   185

13   Saving Lives Through an Accident
Safety Glass   199

14   A Wearable Water Filter
The Sari   217

15   From Ship to Staircase
The History of the Slinky   235

16   Man of Stainless Steel
The Discovery of a New Alloy   253

17   Super Glue
Accidentally Discovered Twice   269

18   Teflon
A Refrigerator Gas Accident   287

19   Vaseline
The "Wonder Jelly"   309

20   I'm Rubber and You're Glue
Vulcanized Rubber   325

21   Peering Into the Unknown
The Discovery of X-Rays   341

22   Velcro
Engineering Mimics Nature   363

Appendix...377

Image Credits...379

Index...383

About the Authors

Dr. M. Gail Jones is an Alumni Distinguished Graduate Professor of Science Education at North Carolina State University. A former middle and high school biology teacher, she leads the Nanoscale Science Education Research Group in investigating effective ways to teach science.

Dr. Elysa Corin is a senior researcher with the Institute for Learning Innovation. She is a former science museum and planetarium educator with experience in astronomy, physics, and engineering education. Her research focuses on the interest-driven and free-choice learning of science.

Megan Ennes is currently a doctoral candidate at North Carolina State University. A former aquarium educator, her research focus includes student science interest and career aspirations as well as the preparation of nonformal educators.

Dr. Emily Cayton is a science education professor at Campbell University who has taught biology and middle school science and has served as an engineering educator. Her research investigates the intersection of policy and practice in science education settings.

Dr. Gina Childers is a professor of middle grades and secondary education at the University of North Georgia. A former middle and high school science teacher, her research explores the use of innovative technologies for teaching science.

Acknowledgments

The authors wish to thank the many people who inspired, reviewed, and helped craft this book. Our special gratitude goes to Sabrina Monserate, Laurel McCarthy, Kendall Rease, and Joseph Gaiteri. We also want to recognize the contributions of Rebecca Hite and thank her for getting us interested in using case studies to teach science. Finally, we thank the teachers who have piloted the book's activities and given us feedback.

Introduction

A number of amazing innovations have resulted from someone making an observation of a phenomenon or trying an experiment just to "see what will happen." For example, George de Mestral noticed how easily cockleburs attached to his pant leg and his dog's fur while mountain hiking in 1948. Curious, he took a closer look at the hook and loop structures of these seeds under a microscope and went on to utilize this unique structure in his trademarked invention of Velcro.

De Mestral addressed a societal need of the 1940s, namely the necessity for a durable clothing fastener. Since this initial work over 70 years ago, Velcro has been used in many applications, from children's shoes to NASA space missions. This example demonstrates that the natural world and results from basic research are filled with ideas and information that may be applied to new products and innovations. The key to harnessing this potential is a careful and imaginative eye, along with a mindful process of engineering to solve everyday problems. This book focuses on this intersection of science and engineering through an examination of real-world discoveries that, as in the case of Velcro, lead to innovations and solutions to contemporary real-world problems. We call the process of creating an innovation from an observation or accidental finding "discovery engineering."

What Is Discovery Engineering?

Discovery engineering begins with the examination of an observation, discovery, or phenomenon. Students review historical observations or discoveries to understand these revelations in their original context. Then they place themselves in the role of the discoverer by thinking about how new insights can be used to create and design new products or applications to solve problems. Authentic details from original studies and data sets make *Discovery Engineering in Physical Science* cases realistic and interesting. Moreover, the case studies use primary documents or historical accounts to engage students in the authentic contexts of science.

With each case study, students explore physical materials, carry out investigations, analyze data, or create models of phenomena before considering further applications for a given innovation. Students are tasked to think creatively about science from serendipity, using research and their own personal insight to create and design new products or applications to solve problems. Throughout the process, students become increasingly knowledgeable about how scientific discoveries often unfold and how engineers apply the design process for creative applications.

Discovery engineering cases engage the learner at multiple levels and scaffold the learning process through observations, an examination of data, and the evaluation and synthesizing of information, followed by an application of the engineering design process to address an everyday problem. Furthermore, the lessons in this book require students to understand and apply fundamental science processes while exploring new ideas for applications.

How Is Discovery Engineering Different From Reverse or Traditional Engineering Design?

Discovery engineering starts with a discovery and is followed by the consideration of an application or a real problem to be solved. As documented in the cases within this book, this is a realistic process that turns traditional "textbook" views of engineering and science upside down. Not all discoveries or inventions are a product of controlled experiments or iterations of engineered prototypes. Some innovations simply begin with an observation, followed by creative thinking and consideration of how that process or phenomenon might be applied in a new way. It is important to note that the case study investigations presented in this book are not designed to teach the traditional approach to engineering (which often starts with a problem). Instead, the investigations start with a student learning about an observation or engineering discovery that was followed by the development of a new product or application. We want students to understand that, just as there is no one way to do science, multiple paths may be taken in engineering. The intent of this book is to teach students that their everyday observations of the world can provide unique insight into the challenges facing modern society. Furthermore, the book seeks to empower students to leverage their natural curiosity to innovate and create new products that alleviate these challenges and foster advancement.

The Case Study Approach

At the heart of each case study in this book is a true story, one that describes how someone made a casual observation or did a simple experiment that led to new insight or a discovery. Case studies are designed to get students actively engaged in the process of problem solving and applying ideas to design new products and processes. The narrative of the case supplies authentic details that help to place the student in the role of inventor and provides scaffolds for critical thinking and deep reflection. A case is more than a paragraph to read or a story to analyze, but rather a way of framing problems, synthesizing information, and thinking creatively about new applications and solutions.

The use of cases as an instructional strategy has had a long history of success in schools of business, law, and medicine. For example, cases are effectively integrated into healthcare-related education programs and used to increase student understanding of the profession, especially for situation-dependent knowledge needed in clinical settings. Cases are an appropriate instructional strategy for the secondary science classroom as they can be used to develop students' critical thinking skills, teach science process skills, and help students think about the nature of science. Additionally, cases are an instructional method that can engender the development of science reasoning skills during non-laboratory classroom time. They guide students to think expertly about problems. They also provide teachers with the opportunity to coach students to use metacognitive strategies as a way to monitor and take control of their own learning. This process reduces rote learning and promotes active engagement. What's more, case studies often enhance student interest by making the topic more relevant to real-life activities.

The case studies provided in this book are designed to supplement instruction by motivating students to apply what they have learned to new contexts and applications. Teaching with case studies provides students with a vicarious experience and casts them in roles that require them to take a different perspective, think differently about science, and take ownership over a decision. This form of instruction encourages students to think critically about a situation; it is valuable as it teaches students to think of many possible solutions to a problem, which approximates the problem-solving environment of many professions in science, technology, engineering, and mathematics (STEM). Cases are especially effective for discussing complex scenarios when there is no single solution to a problem, and they are best integrated into the curriculum when learners can benefit from applying their ideas to a real-world situation.

This book is of value to middle and high school science and engineering teachers as each discovery engineering case includes multiple sections that teachers can tailor to specific classroom environments. Case studies may be used at the start of a unit during the "engage" component of a learning cycle lesson to elicit student interest and provide formative evaluation information about students' preconceptions. A case can also become part of the "extend and apply" component of a lesson. When used at the end of the lesson, the cases help teachers judge whether or not their students understand the science of the case in sufficiently enough to apply their knowledge to new contexts. Case studies contextualize student learning and prompt students to use their knowledge to problem solve in a "real" situation, consider a topic from a new and different perspective, and reflect deeply about their learning. Doing all of this encourages students to increase their understanding of STEM and improve their critical reasoning skills.

Science, Engineering, and the *Next Generation Science Standards*

The *Next Generation Science Standards* (*NGSS*) challenge science teachers to facilitate learning experiences for students that emulate the practices of scientists and engineers (NGSS Lead States 2013). The goal of the *NGSS* is to better situate inquiry in the kinds of work (social, cognitive, and physical) that are authentic to science and engineering. The *NGSS* recommend that K–12 science instruction should do the following:

1. Have broad importance across multiple sciences or engineering disciplines or be a key organizing principle of a single discipline.

2. Provide a key tool for solving problems and understanding or investigating more complex ideas.

4. Relate to the interests and life experiences of students or be connected to societal or personal concerns that require scientific or technological knowledge.

6. Be teachable and learnable over multiple grades at increasing levels of depth and sophistication. That is, the idea can be made accessible to younger students but is broad enough to sustain continued investigation over years (NGSS Lead States 2013, p. xvi).

The case study approach described in this book is designed to help teachers meet these *NGSS* objectives by focusing on real-world problems of interest to the student, employing different levels of depth and sophistication in the problem solving, and integrating thinking across science and engineering.

The *NGSS* specifically state that students should "learn how to engage in engineering design practices to solve problems" (NGSS Lead States 2013, p. 104). Both middle and high school students are expected to know how to define problems, develop solutions, and test and refine their final designs (NGSS Lead States 2013). In middle school, students should be able to define problems precisely and learn how to select the best solution (NGSS Lead States 2013). In high school, students are "expected to engage with major global issues at the interface of science, technology, society and the environment, and to bring to bear the kinds of analytical and strategic thinking that prior training and increased maturity make possible" (NGSS Lead States 2013, p. 128). Accordingly, the discovery engineering cases included here encourage students to use an engineering design model to address current issues with creativity and innovation.

TABLE 1

NGSS Recommendations for Teaching Engineering Practices

Grade Level	ENGINEERING PRACTICES		
	Define	Develop Solutions	Optimize
Early Elementary K–2	Identify situations/ problems that can be solved through engineering	Convey solutions through visual or physical representations	Compare solutions, test, and evaluate
Upper Elementary 3–5	Specify criteria and constraints for a solution to a problem	Research multiple possible solutions	Improve a solution based on results of tests, including failure points
Middle Grades 6–8	Attend to precision, criteria, and constraints that may limit solutions	Combine parts of different solutions to create new solutions	Iteratively test and systematically refine a solution
High School 9–12	Attend to a range of criteria and constraints for problems of social and global significance	Break a problem into smaller problems that can be solved separately	Prioritize criteria, take into account tradeoffs, and assess social and environmental impacts as complex solutions are tested and refined

Source: Adapted from NGSS Lead States 2013, pages 105–106.

Although the *NGSS* does not clearly delineate how teachers are to integrate engineering into science, it makes recommendations about engineering practices for students across grade levels.

The recommendations for teaching engineering practices (Table 1) indicate that students should progress from proposing and testing single solutions to a more complex process of prioritizing and systematically assessing complex solutions to problems.

As students explore engineering and science practices, the *NGSS* raise questions about the classical interpretation of the scientific method. Figure 1 (p. 6) shows the traditional model of the scientific method that has historically been taught in science education. The scientific method is typically represented as beginning with a question. That is followed by an examination of what is known (research), the construction of a hypothesis, the design of an experiment, data collection and analysis, and the drawing of a conclusion. Analysis of experimental results may also lead to forming a new hypothesis and beginning the experimental cycle again The final step involves communicating the result. In actual practice, the methods of science are much more iterative and flexible than are often represented in models of the scientific method. For example, some fields and areas of science (such as geology or astrophysics) do not lend themselves to controlled experiments; rather, advancements in these fields and areas are made through observations and data analysis. The use of engineering cases in science teaching allows teachers to compare science and engineering processes and consider the range of ways both science and engineering can follow or bypass the classical processes presented for these fields.

The engineering design process (Figure 2) involves starting with a problem in need of a solution; imagining a solution (brainstorming ideas); planning a solution (designing diagrams and obtaining materials); creating a product, process, or prototype (following the plan and testing it out); and then improving the design. It is a cyclic process where each iteration leads to a more effective product. Like the scientific method, the engineering design process is often more fluid than many models represent.

Today, it is recommended that science teachers no longer teach the scientific method as a linear process. Rather, teachers are encouraged to teach students problem-solving skills where the domains of science and engineering are blended (NGSS Lead States 2013). The goal is to have students focus on framing questions, developing hypotheses that can be investigated, and engaging in systematic analysis and use of data that can serve as evidence for scientific claims.

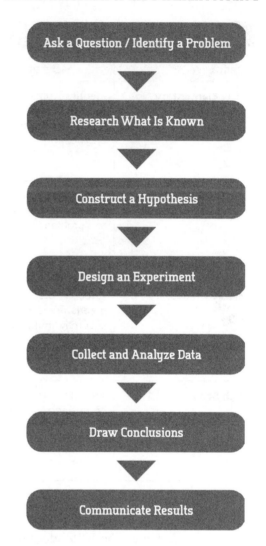

FIGURE 1

Traditional Model of the Scientific Method

Ask a Question / Identify a Problem

Research What Is Known

Construct a Hypothesis

Design an Experiment

Collect and Analyze Data

Draw Conclusions

Communicate Results

Broadening students' understanding of the nature of science is key to teaching students scientific practices and helping them distinguish the work of scientists from that of engineers. It is important for students to develop an understanding of science as a human endeavor that is embedded in previous findings yet open to new interpretations as new evidence is uncovered. The study of engineering also stresses the idea that technologies are driven by human effort and influenced by societal needs and values. Moreover, one of the goals identified in the *NGSS* is for students to understand that "scientists and engineers are guided by habits of mind, such as intellectual honesty, tolerance of ambiguity, skepticism, and openness to new ideas" (NGSS Lead States 2013, p. 69).

The Engineering Design Process

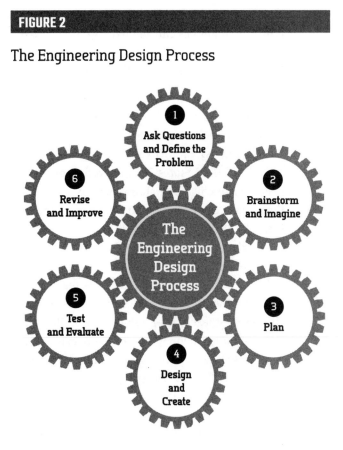

Grand Challenges for Engineering

The discovery engineering cases provide teachers with the opportunity to explain the Grand Challenges for Engineering for the 21st century (National Academy of Engineering 2016). These include the following: advance personalized learning, make solar energy economical, enhance virtual reality, reverse engineer the brain, engineer better medicines, advance health informatics, restore and improve urban infrastructure, secure cyberspace, provide access to clean water, provide energy from fusion, prevent nuclear terror, manage the nitrogen cycle, develop carbon sequestration methods, and engineer the tools of scientific discovery.

The last challenge, engineering the tools of scientific discovery, allows students to think about how advancements in engineering and science occur and consider how new discoveries could solve problems and advance science. For example, students might contemplate advancements in the engineering of nanoparticles, which are now used to package and deliver cancer-fighting drugs directly to cancerous tissues without harming healthy tissues and causing widespread side effects. Furthermore, as students ponder new uses for products such as memory wire, they can consider how new materials can solve the grand challenges that face our society today.

Science and Engineering for All

Discovery engineering cases can help teachers address the needs of every student in the classroom. The *NGSS* highlights the need to serve all students in the science classroom, including underserved groups such as students who are economically disadvantaged, minorities, students with disabilities, and English language learners (NGSS Lead States 2013). "From a pedagogical perspective, the focus on engineering is inclusive of students who may have traditionally been marginalized in the science classroom or experienced science as not being relevant to their lives or future" (NGSS Lead States 2013, p. 104). Researchers have suggested that including engineering practices in the early grades can help keep students interested in engineering and science (Capobianco, Yu, and French 2015).

The cases in these books provide teachers with the tools needed to successfully integrate the engineering design process into their curriculum, no matter what discipline they teach. (See Table 2 on p. 10 for a list of cases by science discipline.) "When students engage with engineering, they are equipped with the capacity to analyze solutions to design problems in their everyday life" (Miller, Januszyk, and Lee 2015, p. 30). As students learn to solve problems, they will be better prepared for the workforce of tomorrow.

Overview of the *Discovery Engineering* Case Study Books

There are two books in this series; the first is organized around physical science concepts, and the second features cases related to biology. Each case in the series is designed for use at either the middle or high school level. Investigations are designed to teach one or more scientific concepts, provide students with background information on the nature of science, and push students to think about new and creative engineering applications. For example, the case in Chapter 2, "Harnessing the Color of Nature: Structural Color and Iridescence" (p. 25), could be used during a lesson on the properties of electromagnetic radiation and color. Showing students a real-world example of how the constructive interference of specific wavelengths of light can produce brilliant colors in bird feathers or butterfly wings may reinforce existing classroom instruction about wave properties, electromagnetic radiation, and the propagation of light. Biology teachers could use the case to promote thinking about animal camouflage and how animals like a cuttlefish use structural color as a way to change their appearance and increase chances of survival. The cases are intended to enhance, illustrate, and extend traditional instruction in ways that promote divergent thinking and encourage students to think like scientists.

Safety Practices in the Science/STEM Laboratory

With hands-on, process, and inquiry-based laboratory activities, the teaching and learning of science today can be both effective and exciting. However, a major challenge to successful engineering instruction revolves around potential safety issues. Teachers can make it safer for students and themselves by adopting, implementing, and enforcing legal safety standards and better professional safety practices in the science classroom and laboratory. This means the inclusion of engineering controls (e.g., ventilation, fume hoods, fire extinguishers, showers), following administrative and safety operating procedures, and using appropriate personal protective equipment (e.g., indirectly vented chemical splash goggles meeting ANSI Z87.1 standard, chemical-resistant nonlatex aprons, nitrile gloves).

Throughout this book, safety notes are provided for laboratory activities and need to be adopted and enforced in an effort to ensure a safer learning/teaching experience. Teachers should also review and follow local polices and protocols used within their school district and/or school (e.g., chemical hygiene plans, Board of Education safety policies).

The National Science Teachers Association (NSTA) provides a list of science rules and regulations, including standard operating procedures for lab safety and a safety acknowledgment form for students and parents or guardians to sign. You can access these resources at *http://static.nsta.org/pdfs/SafetyInTheScienceClassroom.pdf*.

Disclaimer: The safety precautions of each activity are based in part on use of the recommended materials and instructions, legal safety standards, and better professional practices. Selection of alternative materials or procedures for these activities may jeopardize the level of safety and therefore is at the user's own risk.

References

Capobianco, B., J. Yu, and B. French. 2015. Effects of engineering design-based science on elementary school science students' engineering identity development across gender and grade. *Research in Science Education* 45 (2): 275–292.

Dowd, S. B., and R. Davidhizar. 1999. Using case studies to teach clinical problem-solving. *Nurse Educator* 24 (5): 42–46.

Gallucci, K. 2006. Learning concepts with cases. *Journal of College Science Teaching* 36 (2): 16–20.

Miller, E. C., R. Januszyk, and O. Lee. 2015. Engineering progressions in the *NGSS* diversity and equity case studies. *Science Scope* 38 (9): 27–30.

National Academy of Engineering. 2016. *Grand challenges for engineering: Imperatives, prospects, and priorities.* Washington, DC: National Academies Press.

NGSS Lead States. 2013. *Next Generation Science Standards: For states, by states.* Washington, DC: National Academies Press.

Smith, R. A., and S. K. Murphy. 1998. Using case studies to increase learning and interest in biology. *American Biology Teacher* 60 (4): 265–268.

TABLE 2

Discovery Engineering Cases by Science Discipline

Case	SCIENCE DISCIPLINE			
	Chemistry	Physics	Earth/Space	Biology
Don't Get Bent Out of Shape: Memory Wire	●			●
Harnessing the Color of Nature: Structural Color and Iridescence		●		●
From Sludge to Nobility: Artificial Dye	●			
By the Teeth of Your Skin: Shark Skin and Bacteria		●		●
A Sticky Situation: Gecko Feet Adhesives		●	●	
It's Hip to Be Square: Seahorse Tails		●		●
Corn Flakes: Waste Not, Want Not	●			
Get Fired Up With Friction Lights: Matches	●	●		
That's a Wrap: Plastic	●			
A Sticky Discovery: The Invention of Post-It Notes	●			
Putty in Your Hands: The Unintended Discovery of Silly Putty	●		●	
Sweet and Sour: The Rise and Fall ... and Rise ... of Saccharin	●			
Saving Lives Through an Accident: Safety Glass	●	●		
A Wearable Water Filter: The Sari		●		●
From Ship to Staircase: The History of the Slinky		●		
Man of Stainless Steel: The Discovery of a New Alloy	●	●		
Super Glue: Accidentally Discovered Twice	●			
Teflon: A Refrigerator Gas Accident	●	●		
Vaseline: The "Wonder Jelly"	●			
I'm Rubber and You're Glue: Vulcanized Rubber	●	●		
Peering Into the Unknown: The Discovery of X-Rays		●	●	●
Velcro: Engineering Mimics Nature		●		●

DON'T GET BENT OUT OF SHAPE

Memory Wire

A Case Study Using the Discovery Engineering Process

Introduction

Imagine accidentally crushing the frames of your glasses or denting your car—and being able to just reheat these objects back into their original shapes! This can actually be done with an innovative material called *memory wire*. After the wire is set at high temperature to a particular shape, it can be bent at room temperature into another shape. (See Figure 1.1, p. 12.) When heated again with just a little heat (as from a hair dryer), the metal will "remember" its initial shape and spring back into that original form. Memory wire, also known as Nitinol wire, has been around for 50 years and was first used by the U.S. military. The material is lightweight and flexible, and it plays a role in many medical, automotive, robotics, and space applications in addition to ordinary consumer products.

Lesson Objectives

By the end of this case study, you will be able to

- Describe the physical and chemical characteristics of memory wire.

- Analyze the benefits and limitations of memory wire as a material for consumer products.

- Design a new application for memory wire that solves a problem.

The Case

Read the following summary of the article "Accidents can happen—in a very good way," published in the *Fresno Bee* on June 25, 2005. Featuring an interview with the inventor of memory wire, the article describes how advancements in science and engineering are often unexpected yet lead to valuable applications. Once you finish reading, answer the questions that follow.

William Buehler, a U.S. Naval Ordnance Laboratory metallurgist (or scientist who studies metals), recounted the story behind the discovery of memory wire. In 1958, Buehler was seeking an impact- and heat-resistant alloy with low density for a missile nose cone that could better withstand the heat created as a missile passes through Earth's atmosphere. The material also had to be fatigue resistant. Metal fatigue is the weakening of metal due to stress, which results in cracks in the metal (see Figure 1.2). While testing various alloys, Buehler noticed that a nickel-titanium alloy was distinctly different from the others. It possessed the desired properties and could easily be formed into a wire. Buehler named his discovery Nitinol (a combination of the terms *nickel, titanium,* and *naval ordnance laboratory*).

At a management meeting to demonstrate Nitinol's resistance to fatigue, Buehler's assistant pulled out one of their "props": an accordion-folded strip of the alloy. It was passed out around the table and flexed by all present. Technical director David S. Muzzey applied heat from his pipe lighter to the compressed Nitinol strip. To everyone's amazement, it stretched out to its original form. This was the first demonstration of the shape-memory capability of Nitinol and the first of the so-called smart or intelligent metals, which are characterized as metals that "think for themselves," sense changes in themselves and their environments, and respond appropriately.

Recognize, Recall, and Reflect

1. What was William Buehler originally looking for? Why?

2. What materials is Nitinol made of?

3. What happened at the meeting that revealed Nitinol's unique shape-changing property?

FIGURE 1.1

Memory Wire

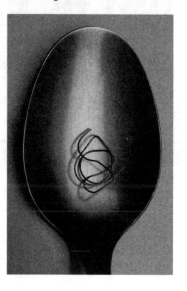

Investigate

In this activity, you will explore the unique features of memory wire. First, you will investigate how memory wire behaves through active exploration (inquiry) and then you will design your own experiment using memory wire.

Materials

- 1 piece of memory wire per student

- Hot water (not boiling)

- Pyrex bowl or borosilicate glass bowl; don't use plastic (groups of students can share bowls)

- Metal tongs (groups of students can share tongs)

- Indirectly vented chemical splash safety goggles (1 pair per student)

- Nonlatex apron (1 per student)

- Heat-resistant gloves (1 pair per student)

FIGURE 1.2

Metal Fatigue on an iPhone Charger

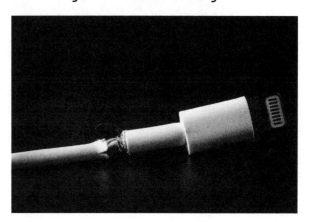

Safety Note: Chemical splash safety goggles, a nonlatex apron, and heat-resistant gloves must be worn during the setup, hands-on, and takedown segments of the activity. The water is very hot and tongs should be used to handle the memory wire. Use caution when using sharp objects, which can cut or puncture skin. Immediately wipe up any spilled water on the floor as it is a slip/fall hazard. Wash your hands with soap and water immediately after completing this activity.

Create, Innovate, and Investigate

- Begin by examining a piece of memory wire. What do you observe about the wire? Is it different from other wires that you have seen?

- Bend the wire into different shapes. Does it have any unusual properties?

- Bend the wire into a corkscrew or a heart shape. Using tongs, drop the wire into a bowl of hot water. What do you observe?

- Remove the wire from the water with the tongs and place it on the table for a few seconds. Observe the wire now that it has been heated. Does it look the same as it did before you heated it?

- Repeat the investigation with different shapes and water of different temperatures (i.e., hot and cold). What can you conclude about the behavior of the memory wire?

Questions for Reflection

1. What did you observe about the physical properties of memory wire?

2. What is needed to make the metal "remember" its original shape?

2. When you heat this metal, how does it behave differently from when you heat other materials like water?

Apply and Analyze

Memory wire is quickly being adopted for new areas of research and new commercial applications. So, how does this material work? The nickel-titanium alloy that William Buehler invented gets its unusual properties from its atomic structure. Buehler noticed while experimenting with the alloy that striking the metal when it was cold created a dull thud. But when he struck the metal at a higher temperature, the sound was more similar to that of a bell. This was an important clue: Buehler knew that a metal's atomic structure affects how vibrations move through it and influences the sounds it will produce. This meant that the atomic structure of the metal had changed as it changed temperature!

Memory metals have two different phases in the solid state, each with two different arrangements of atoms. When memory wire is heated to around 500°C, it forms strong, fixed bonds between atoms, similar to steel. This is the high-temperature solid phase. While in the lower-temperature solid phase (around room temperature), the bonds between the atoms in the memory wire are weaker and the atoms can slip past one another to make for a more flexible material.

At lower temperatures the material can be easily bent into a certain shape. When the temperature is then raised to 500°C, the metal sets, and it will "remember" that shape. After the metal is set, you can lower the temperature and bend the material in a different way. When a small amount of heat is then added to the metal, it will pass through its phase change, transitioning from the low-temperature solid state to the high-temperature solid state. During the transition, the metal will rapidly change back to its original form.

Memory metal is being used in a wide range of products—from dental braces that tighten around teeth when heated to tools used to open clogged arteries.

Watch the video "Magic metals, how shape memory alloys work" from Ted-Ed for an animated explanation of this phase change: *www.youtube.com/watch?v=yR-6_lS9vts*. Then explore each web link listed below and answer the questions that follow.

- *http://iopscience.iop.org/0964-1726/16/6/060*

- *www.hcltech.com/sites/default/files/design-engineering-of-Nitinol.pdf*

1. What are some of the cost-versus-benefit concerns of using memory wire?

2. Does memory wire last forever? Are there limitations to using memory wire? What happens to it over time?

4. What are some new applications for memory wire that are being explored?

Design Challenge

Engineering is the application of scientific understanding through creativity, imagination, problem solving, and the designing and building of new materials to address and solve problems in the real world. You will be asked to take the science you have learned in this case and design a process or product to address a real-world issue of your choosing.

Engineers use the engineering design process as steps to address real-world problems (see Figure 1.3, p. 16). You will now use this process as you come up with a new way to use memory wire. In this case, you are asking the question (Step 1) of how you can design a new use for memory wire. Drawing on your creativity, you will then brainstorm (Step 2) a new product that uses memory wire to solve a problem. Afterward, you will create a plan (Step 3) for this new product. Next, you will create a sketch and/or model of your product (Step 4). Then, you will work with your classmates to think about how you would test (Step 5) and refine (Step 6) your product.

1. Ask Questions

Based on your previous research, consider a new problem that may be addressed or a product that could be created using a metal that "remembers" its shape. What are some applications in which you would need a material that could return to its initial form after being bent?

2. Brainstorm and Imagine

Memory metal is being used in many areas including science, business, and health. Go to the following link to read about uses for shape memory metal: *https://depts. washington.edu/matseed/mse_resources/Webpage/Memory%20metals/applications_for_shape_memory_al.htm*. After you're done, begin to brainstorm a specific new application for memory wire. (For example, many elderly people with arthritis in their fingers have trouble using buttons on shirts. Perhaps a button made of memory wire could be made that is shaped like a cone to easily move through buttonholes. And with the addition of heat from a hair dryer, it could pop back into a button shape. Another idea is to use memory wire to make the body of a car. Then if the car is hit and dented, you could use heat to pop the car back into shape.)

3. Create a Plan

Create a plan for your product. Consider: (1) What is the purpose of the memory wire product? (2) What are the benefits of the product? (3) What are the limitations of the product? (Are there only some ways memory wire can be used or only some conditions where it works?) Use the Product Planning Graphic Organizer (p. 18) to help you.

(p. 18)

FIGURE 1.3

The Engineering Design Process

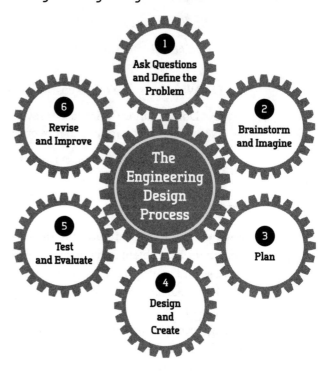

4. Design and Create

Consider the following questions and considerations for your memory wire–based product and its design.

- How would incorporating memory wire make this product better?

- Are there any limitations or drawbacks to using memory wire? If so, how would you overcome them?

- What technologies might need to be developed to create or manufacture this design?

- What are any limiting factors you can foresee with implementing this design?

- Would there be any safety concerns regarding your memory wire–based product?

Now create a sketch of your memory wire design. Make sure your design incorporates the research and exploration you've done.

5. Test and Evaluate

Working with your classmates, come up with a way to test your design to see its effectiveness.

6. Revise and Improve

Give your plans to one of your classmates for review. Listen to his or her feedback on your design and take some time to revise and make improvements. What are some ways you can use the input to refine your design?

Reflect

1. What technologies might need to be developed to create or manufacture this design?

2. What are any constraints or drawbacks you can foresee with implementing this design?

4. Would there be any environmental or human health concerns about your design?

Product Planning Graphic Organizer

Proposed Product Idea	
Pros (Benefits)	**Cons (Limitations)**

TEACHER NOTES

DON'T GET BENT OUT OF SHAPE
MEMORY WIRE

A Case Study Using the Discovery Engineering Process

Lesson Overview

In this lesson, students explore a metal that can be bent and, when lightly heated, "remembers" and springs back into its original shape. Memory metal was an accidental discovery that has led to a number of applications and products.

Lesson Objectives

By the end of this case study, students will be able to

- Describe the physical and chemical characteristics of memory wire.

- Analyze the benefits and limitations of memory wire as a material for consumer products.

- Design a new application for memory wire that solves a problem.

The Case Study Approach

This lesson uses a case study approach. Explaining the purpose of case studies will encourage your students to relate to the material and engage with the problem. At the heart of each case study in this book is a true story, one that describes how someone in his or her everyday life or during a routine workday made an observation or did a simple experiment that led to a new insight or discovery. Case studies are designed to get students actively engaged in the process of problem solving. The narrative of the case supplies authentic details that place the student in the role of the inventor and provide scaffolds for critical thinking and deep reflection. A case is more than a paragraph to read or a story to analyze but rather a way of framing problems, synthesizing what is known, and thinking creatively about new applications and solutions. In this lesson, students consider how memory wire was discovered and work together to think about new applications for memory wire that solve real-life problems.

Use of the Case

Due to the nature of these case studies, teachers may elect to use any section of each case for their instructional needs. The sections are sequenced in order (scaffolded) so students think more deeply about the science involved in the case and develop an understanding of engineering in the context of science.

Curriculum Connections

Lesson Integration

You could use this case as a way to integrate engineering design into a lesson on atomic structure or a discussion of metals and the periodic table.

Related Next Generation Science Standards

PERFORMANCE EXPECTATIONS

- HS-PS2-6. Communicate scientific and technical information about why the molecular-level structure is important in the functioning of designed materials.

- HS-ETS1-3. Evaluate a solution to a complex real-world problem based on prioritized criteria and trade-offs that account for a range of constraints, including cost, safety, reliability, and aesthetics, as well as possible social, cultural, and environmental impacts.

SCIENCE AND ENGINEERING PRACTICES

- Analyzing and Interpreting Data

- Engaging in Argument From Evidence

- Constructing Explanations and Designing Solutions

CROSSCUTTING CONCEPT

- Structure and Function

Related National Academy of Engineering Grand Challenges

- Engineer Better Medicines

- Advance Health Informatics

- Engineer the Tools of Scientific Discovery

Lesson Preparation

You will need to make copies of the entire student section for the class. Students will need internet access at various points in the lesson. Alternatively, you can project videos or print and distribute copies of online content for the class. For the Investigation section, you will need hot water ready for the investigation and tongs that you can use to retrieve the memory wire after it is immersed. Memory wire can be purchased from a variety of sources, including Carolina Biological Supply, *teachersource.com*, and Educational Innovations. It is helpful to buy it by the foot and then cut it into 4- to 6-inch pieces. Look at the Teaching Organizer (Table 1.1, p. 22) for suggestions on how to organize the lesson.

Materials

- 1 piece of memory wire per student

- Hot water (not boiling)

- Pyrex bowl or borosilicate glass bowl; don't use plastic (groups of students can share bowls)

- Metal tongs (groups of students can share tongs)

- Indirectly vented chemical splash safety goggles (1 pair per student)

- Nonlatex apron (1 per student)

- Heat-resistant gloves (1 pair per student)

Safety Note for Students: Chemical splash safety goggles, a nonlatex apron, and heat-resistant gloves must be worn during the setup, hands-on, and takedown segments of the activity. The water is very hot and tongs should be used to handle the memory wire. Use caution when using sharp objects, which can cut or puncture skin. Immediately wipe up any spilled water on the floor as it is a slip/fall hazard. Wash your hands with soap and water immediately after completing this activity.

Time Needed

55 minutes

TABLE 1.1

Teaching Organizer

Section	Time Suggested	Materials Needed	Additional Considerations
The Case	5 minutes	Student packet	Could be read in class or as a homework assignment prior to class
Investigate	10 minutes	Student packet, 1 piece of memory wire, hot water, Pyrex bowl, tongs, indirectly vented chemical splash safety goggles, nonlatex apron, heat-resistant gloves	Recommended as an in-class demonstration with students individually testing their memory wire. It helps for the teacher to control the pouring of hot water into the bowls.
Apply and Analyze	10 minutes	Student packet, internet access	Whole-class or individual activity
Design Challenge	30 minutes	Student packet, internet access	Small-group activity

Teacher Background Information

Students often have questions about metals and atomic structure, so it may be useful to review the properties of metals. There are a number of resources and videos about shape memory materials available on the internet. You may want to observe the behavior of memory wire on sites such as YouTube (see, for example, "SpaceMETA present NITINOL explanation by MIT" at *www.youtube.com/watch?v=2YVwpBAiA1A*) prior to using the case.

Vocabulary

- alloy
- atom
- crystal
- metal

- metallurgist
- Nitinol
- phase change

Teacher Answer Key

Recognize, Recall, and Reflect

1. **What was William Buehler originally looking for? Why?**

   He was looking for an impact- and heat-resistant alloy with low density for a missile nose cone.

2. **What materials is Nitinol made of?**

   Nickel and titanium

3. **What happened at the meeting that revealed Nitinol's unique shape-changing property?**

   Technical director David S. Muzzey applied heat from his pipe lighter to the compressed Nitinol strip, and it changed shape.

Questions for Reflection

1. **What did you observe about the physical properties of memory wire?**

   The metal can be bent and then heated to transform back into its original shape.

2. **What is needed to make the metal "remember" its original shape?**

   The metal must be heated slightly to return to its original shape.

3. **When you heat this metal how does it behave differently from when you heat other materials like water?**

   Other materials like water change state when heated. Memory wire does not change state unless exposed to an extremely high temperature.

Apply and Analyze

1. **What are some of the cost-versus-benefit concerns of using memory wire?**

   The metal is relatively inexpensive so the benefits are worth the costs for applications like eye glasses repair. However, it's too costly to be used in the large-scale production of items such as paperclips.

2. **Does memory wire last forever? What happens to it over time?**

   Eventually memory wire does experience structural fatigue and must be replaced if used frequently in an application.

3. **What are some new applications for memory wire that are being explored?**

Robotics, telecommunications, optometry, dentistry, and orthopedic surgery

Reflect

1. **What technologies might need to be developed to create or manufacture this design?**

Facilities to heat the metal to high temperatures to set the shape.

2. **What are any constraints or drawbacks you can foresee with implementing this design?**

Any application that uses memory wire would need to account for metal fatigue.

3. **Would there be any environmental or human health concerns?**

Answers will vary.

Assessment

The Design Challenge can be assessed using the rubric in the appendix (p. 377).

Extensions

This lesson can be followed with lessons on plasticity of materials, properties of metals, and metal fatigue.

Resources and References

Brown, T. L., H. E. LeMay, B. E. Bursten, and J. R. Burdge. 2003. *Chemistry: The central science.* 9th ed. Upper Saddle River, NJ: Pearson Education, Inc.

Gopal, V., and G. Mani. Design engineering of Nitinol-based medical devices. HCL Technologies. *www.hcltech.com/sites/default/files/design-engineering-of-Nitinol.pdf.*

Images Scientific Instruments. Nitinol-shape memory materials: Introduction. *www.imagesco.com/articles/Nitinol/01.html* (accessed June 16, 2016).

Pappas, P. et al. 2007. Transformation fatigue and stress relaxation of shape memory alloy wires. *Smart Materials and Structures* 16 (6): 2560–2670.

SpaceMETA. 2013. "SpaceMETA present NITINOL explanation by MIT." YouTube video. *www.youtube.com/watch?v=2YVwpBAiA1A.*

Swardz, S. Applications for shape memory alloys. University of Washington. *https://depts.washington.edu/matseed/mse_resources/Webpage/Memory%20metals/applications_for_shape_memory_al.htm* (accessed February 5, 2019).

Ted-Ed. 2012. "Magical metals, how shape memory alloys work." YouTube video. *www.youtube.com/watch?v=yR-6_lS9vts.*

Woodford, C. 2010. Shape-memory materials. *www.explainthatstuff.com/how-shape-memory-works.html.*

HARNESSING THE COLOR OF NATURE

Structural Color and Iridescence

> ### A Case Study Using the Discovery Engineering Process

Introduction

Have you ever noticed that certain butterflies, birds, and beetles boast colors that aren't just bright and brilliant but actually shine (see Figure 2.1, p. 26)? These creatures' colors are iridescent: The colors shift and change slightly as the creature moves and is viewed from different angles. Plants, animals, and other objects that present iridescent colors have long been admired for their beauty. However, only in the past few decades have scientists come to understand that this iridescent coloring is not due to pigments, molecules that give color to many plants and animals (including humans) by absorbing and reflecting specific colors of light. Rather it's the result of another coloring process entirely. Read on to understand what causes this brilliant coloring and how scientists and engineers may use the principles involved to create new products.

Lesson Objectives

By the end of this case study, you will be able to

- Describe why materials that get their color from their nanoscale surface structures look different from those that are colored by pigments.

- Explain how nanoscale surface structures can create iridescent colors.

- Invent a new product that uses structural color to solve a problem.

FIGURE 2.1

Iridescent Insects

The blue morpho butterfly (*Morpho rhetenor*) at left is iridescent and gets its color from nanoscale structures on its scales. The monarch butterfly (*Danaus plexippus*) at right is not iridescent and gets its orange color from pigments.

The Case

Read the following case summarized from the article "Color From Structure" by Cristina Luiggi, which was published in *The Scientist* magazine in February 2013. The full article begins with the statement, "Colors may be evolution's most beautiful accident." It goes on to describe how, when, and why some plant and animal species may have evolved to have structural color. It also explains how nanoscale surface structures give rise to these brilliant colors.

While working as a physicist at the University of Cambridge, Dr. Silvia Vignolini's work on the behavior of light took an unexpected turn when she visited England's Royal Botanic Gardens, Kew. At the gardens, she came across an unusual berry from the African plant *Pollia condensata* (see Figure 2.2). The berries were dazzling in appearance: a potent blue color that looked a little metallic. Vignolini was surprised to learn that the vivid samples she was looking at had been collected more than forty years before! She thought it was odd that the fruit produced a color that wouldn't fade for so many years. Vignolini decided to figure out how the berry was able to shine with such an intense iridescent color that lasted decades after it was picked.

FIGURE 2.2

The Berry of the *Pollia Condensata* Plant

First, Vignolini and her colleagues tried to isolate blue pigments from the fruit in order to examine them but were unable to do so. Next, the scientists looked at the berries with electron microscopes. These powerful microscopes can detect features as small as a nanometer, which is one *billionth* of a meter. Using the microscopes the scientists observed that the smooth surface of the berry wasn't smooth at all. On the berry's surface were structures too small to see with the human eye. They were made of cellulose and stacked in a helix pattern like a spiral staircase. The scientists determined that the height of one complete "turn" of the spiral is similar in size to the wavelength of blue light and that the size and shape of the cellulose helix causes the berry to only reflect blue light, giving it its blue color. If the cellulose structures and empty spaces in the pattern were a little larger or smaller, the berry would reflect a different color of light. Colors that are caused by the nanoscale structure of a material, as in the case of the *Pollia condensata* berry, are called structural colors. The berry Vignolini had examined that was picked many years before had not faded in color because the nanoscale surface structures are very stable. This is different from pigment molecules, which may degrade and change over time, causing their colors to fade. Structural colors can last for millions of years.

Recognize, Recall, and Reflect

1. What was the surprising observation Silvia Vignolini made about the *Pollia condensata* berry that prompted her to learn more about it?

2. In your own words, what causes the blue iridescence of the *Pollia condensata* berry?

3. What property of the berry has to change in order to show other colors such as red, purple, or green?

Investigate

Have you ever looked at the swirling colors in a soap bubble and wondered how a clear object could be colorful? A soap bubble is a layer of soapy water surrounding an air pocket. The layer of soap is very thin—approximately hundreds of nanometers thick. The thickness of the soap wall is always changing as the liquid swirls around the bubble, with gravity pulling more and more soap toward the bottom of the bubble over time. Sunlight contains all of the colors of the visible light spectrum and shines on the bubble. Depending on the thickness of the bubble at a specific location, you will observe some colors and not others in the sunlight. You can even observe a bubble shift color as the soap moves around the bubble layer, changing its thickness. This is due to thin film interference, a process by which the thickness of the bubble at a specific location determines which colors will shine brightest.

Thin film interference is a very simple example of structural color: A thin, flat layer of a transparent material interacts with white light and only reflects certain colors depending on the thickness of the thin film layer. In this activity, you will create your own object that gets its color from structure. Your final product will produce a range of iridescent colors—from red to violet—due to the constructive and destructive interference of light.

Materials

For each individual or group of students:

- Container filled with at least 1 in. of water

- Additional water

- Strips of black construction paper

- 1 bottle of clear nail polish

- Paper towels

- Indirectly vented chemical splash safety goggles (1 pair per student)

- Nonlatex apron (1 per student)

- Nitrile gloves (1 pair per student)

Safety Note: Carry out this activity in a well-ventilated room or under a fume hood and return the cap to the nail polish bottle when you are not using it. Wear indirectly vented chemical splash safety goggles, a nonlatex apron, and nitrile gloves during the setup, hands-on, and takedown segments of the activity. Immediately wipe up any spilled water on the floor as it is a slip/fall hazard. Wash your hands with soap and water immediately after completing this activity.

Create, Innovate, and Investigate

- Completely submerge one strip of black paper underwater, holding it down with one of your hands.

- With your other hand, dip the nail polish brush into the clear nail polish. Hold the brush with polish over the surface of the water.

- Wait a few seconds for the clear nail polish to drip onto the surface of the water. Watch it spread out in a thin layer over the water's surface. (*Note:* Do not touch the tip of the nail polish brush to the water. Wait for the nail polish to drip off on its own.)

- Pull the submerged piece of black paper up through the clear layer of nail polish.

- Set the black paper strip on a piece of paper towel to dry. You have successfully made an artwork that displays iridescent colors through thin film interference!

- If you are making multiple film-covered strips of paper, use your gloved finger or a paper towel to collect leftover nail polish film in the water. Then repeat the steps from the beginning. You may eventually need to change the water in your dish. You will know that it is time to change the water when a nail polish droplet that's introduced doesn't spread out into a film on the surface of the water.

- Once the thin film has dried on your black paper strip (which takes about 10 minutes), answer the questions below.

Questions for Reflection

1. What colors do you observe in the film-covered strip of paper? What happens when you move the strip around and look at it from different angles?

2. You applied a thin layer of clear nail polish to black paper. In your own words, explain how this process has created the colors you are observing.

3. If you wanted to create a thin film that was all the same color (green, for example), what would need to be different about your thin film?

Apply and Analyze

Over 500 million years ago, an event known as the Cambrian explosion caused a surge in the diversity of creatures living on Earth. It was around this time that a random mutation in an organism's genetic code made a small change that led to the development of structural colors. Life on Earth continued to evolve and diversify, and there are now many different nanoscale shapes and structures that produce colors. Below are a few examples.

- As you read above, the berry of the *Pollia condensata* plant has developed a surface structure made up of nanoscale stacks of cellulose. Each layer is slightly rotated in position compared to the layer in the stack below it, similar to a spiral staircase. The height of each turn in the staircase determines which color of visible light is reflected, most often a shade of blue.

- The wing scales of the blue morpho butterfly (*Morpho rhetenor*) are covered with chitin nanostructures that look like evergreen trees as they feature a pole with several layers of parallel branches. These "trees" point outward from the surface of the wing, and the spacing between the branches is the right distance apart to reflect blue light.

- A male Lawes's parotia (*Parotia lawesii*) bird (see Figure 2.3), has barblike structures on its breast feathers. The barbs contain layers of melanin whose spacing reflect a yellow-orange color. These barbs also have a V-shaped component that can reflect blue light. As the bird moves, his feathers reflect different wavelengths of light toward the observer and a color change from yellow to green to blue may be observed.

FIGURE 2.3

Male and Female Lawes's Parotia Birds of Paradise

Although the specific shapes of these nanostructures might differ from organism to organism, they all manipulate light in specific ways to produce structural color. That is to say, the structural color of each of these organisms is created by an array of nanoscale surface features whose size and pattern allow light of certain colors (wavelengths) to be reflected. Sometimes these colors can even appear to shift—greens turn to blues, oranges turn to reds—as the organism moves relative to the observer.

Our knowledge of structural color isn't just useful in understanding today's organisms. It also gives us insight into living things of the past. By combining this information with observations through powerful microscopes of exceptionally well-preserved fossils, we can determine the likely color of extinct creatures.

Watch the TED-Ed video "How do we know what color dinosaurs were?" (*http://ed.ted.com/lessons/how-do-we-know-what-color-dinosaurs-were-len-bloch*), which explains how scientists use fossils and their knowledge of structural color to determine the color of certain dinosaurs. When you are done, answer the following questions.

1. When did the *Microraptor* live?

2. What clues do scientists look for in a fossil when they are trying to determine the color of an extinct animal?

3. Can scientists use this technique to determine the color of all extinct animals from fossils?

Design Challenge

Engineering is the application of scientific understanding through creativity, imagination, problem solving, and the designing and building of new materials to address and solve problems in the real world. You will be asked to take the science you have learned in this case and design a process or product to address a real-world issue of your choosing.

FIGURE 2.4

The Engineering Design Process

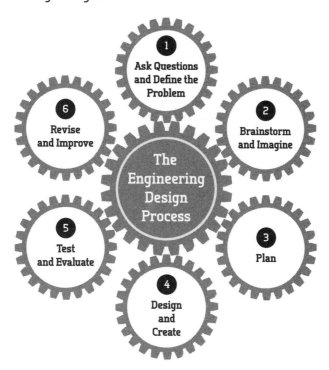

Engineers use the engineering design process as steps to address real-world problems (see Figure 2.4). You will now use this process as you come up with a new way to use structural color. In this case, you are asking the question (Step 1) of how you can design a new use for a structural color. Drawing on your creativity, you will then brainstorm (Step 2) a new product that uses structural color to solve a problem. Afterward, you will create a plan (Step 3) for this new product. Next, you will create a sketch and/or model of your product (Step 4). Then, you will work with your classmates to think about how you would test (Step 5) and refine (Step 6) your product.

1. Ask Questions

Based on your research above, consider a new problem that may be addressed or a product that could be created by using structural color. What are some applications in which you might need colors that don't easily fade and can change when viewed from different angles?

2. Brainstorm and Imagine

Use what you have learned about structural color to brainstorm a product that gets its bright, long-lasting color from its nanostructure rather than from dyes or pigments. (For example, a shirt that uses this technology would remain colorful over time and, therefore, might not need to be replaced as soon as one that is dyed.) What problem would your product solve?

3. Create a Plan

Create a plan for your new product. Consider: (1) What is the purpose of the product? (2) What are the benefits of the product? (3) What are the limitations of the product? Use the Product Planning Graphic Organizer (p. 34) to help you.

4. Design and Create

Consider the following questions and considerations for your structural color product and its design.

- How would incorporating structural color make this product better?

- Are there any limitations or drawbacks to using structural color in this product? If so, how would you overcome them?

- What technologies might need to be developed to create or manufacture this design?

- What are any constraints or drawbacks you can foresee with implementing this design?

- Would there be any safety concerns regarding your product?

Now create a sketch of your structural color design. Make sure your design incorporates the research and exploration you've done.

5. Test and Evaluate

Working with your classmates, come up with a way to test your design to see its effectiveness.

6. Revise and Improve

Give your plans to one of your classmates for review. Listen to his or her feedback on your design and take some time to revise and make improvements. What are some ways you can use the input and refine your design?

Reflect

1. What technologies might need to be developed to create or manufacture this design?

2. What are any constraints or drawbacks you can foresee with implementing this design?

3. Would there be any environmental or human health concerns about using structural color in this way?

Product Planning Graphic Organizer

Proposed Product Idea

Pros (Benefits)	Cons (Limitations)

HARNESSING THE COLOR OF NATURE

STRUCTURAL COLOR AND IRIDESCENCE

A Case Study Using the Discovery Engineering Process

Lesson Overview

In this lesson, students explore how certain organisms, including some plant and animal species, display beautiful iridescent colors when light interacts with nanoscale structures on their surfaces. Students will make an object to take home that creates iridescent colors through thin film interference, a simple mechanism for producing structural color. This activity was chosen to illustrate the information in the case study because it presents students with a perplexing situation: How can a material with no color (in this case clear nail polish) create the range of colors observed in the final product? This surprising result further reinforces the fact that the color students observe is not due to a pigment but rather another process. Students will also read about the more complex nanoscale structures that living organisms use to produce color and how knowledge of structural color is being used by scientists to learn more about extinct creatures.

Lesson Objectives

By the end of this case study, students will be able to

- Describe why materials that get their color from their nanoscale surface structures look different from those that are colored by pigments.

- Explain how nanoscale surface structures can create iridescent colors.

- Invent a new product that uses structural color to solve a problem.

The Case Study Approach

This lesson uses a case study approach. Explaining the purpose of case studies will encourage your students to relate to the material and engage with the problem. At the heart of each case study in this book is a true story, one that describes how someone in his or her everyday life or during a routine workday made an observation or did a simple experiment that led to a new insight or discovery. Case studies are designed to get students actively engaged in the process of problem solving. The narrative of the case supplies authentic details that place the student in the role of the inventor and provide scaffolds for critical thinking and deep reflection. A case is more than a paragraph to read or a story to analyze but rather a way of framing problems, synthesizing what is known, and thinking creatively about new applications and solutions. In this lesson, students consider how structural color was discovered and work together to think about new applications for structural color that solve real-life problems.

Use of the Case

Due to the nature of these case studies, teachers may elect to use any section of each case for their instructional needs. The sections are sequenced in order (scaffolded) so students think more deeply about the science involved in the case and develop an understanding of engineering in the context of science.

Curriculum Connections

Lesson Integration

You could use this case as a way to integrate engineering design into a lesson on electromagnetic radiation or the properties of waves. It would be especially appropriate after a lesson on constructive and destructive interference. The case would also work well with Chapter 3, "From Sludge to Nobility: Artificial Dye" (p. 43).

Related Next Generation Science Standards
PERFORMANCE EXPECTATIONS

- MS-PS4-2. Develop and use a model to describe that waves are reflected, absorbed, or transmitted through various materials.

- MS-LS4-2. Apply scientific ideas to construct an explanation for the anatomical similarities and differences among modern organisms and between modern and fossil organisms to infer evolutionary relationships.

- HS-PS4-1. Use mathematical representations to support a claim regarding relationships among the frequency, wavelength, and speed of waves traveling in various media.

SCIENCE AND ENGINEERING PRACTICES

- Engaging in Argument From Evidence
- Constructing Explanations and Designing Solutions

CROSSCUTTING CONCEPTS

- Structure and Function
- Cause and Effect
- Scale Proportion and Quantity

Related National Academy of Engineering Grand Challenge

- Engineer the Tools of Scientific Discovery

Lesson Preparation

You will need to make copies of the entire student section for the class. Students will need internet access at various points in the lesson. Alternatively, you can project videos or print and distribute copies of online content for the class. For the Investigate section, cut the construction paper into strips small enough to be completely submerged into your water container. Strips approximately 2 × 4 inches work well. Depending on how many film-covered strips the students make, they will need access to water to change out the water in their containers. Look at the Teaching Organizer (Table 2.1, p. 38) for tips on how to organize the lesson.

Materials

For each individual or group of students:

- Container filled with water, at least 1 in. deep
- Additional water
- Strips of black construction paper
- 1 bottle of clear nail polish
- Paper towels
- Indirectly vented chemical splash safety goggles (1 pair per student)
- Nonlatex apron (1 per student)
- Nitrile gloves (1 pair per student)

Safety Note for Students: Carry out this activity in a well-ventilated room or under a fume hood and return the cap to the nail polish bottle when you are not using it. Wear indirectly vented chemical splash safety goggles, a nonlatex apron, and nitrile gloves during the setup, hands-on, and takedown segments of the activity.

Immediately wipe up any spilled water on the floor as it is a slip/fall hazard. Wash your hands with soap and water immediately after completing this activity.

Time Needed

75 minutes

TABLE 2.1

Teaching Organizer

Section	Time Suggested	Materials Needed	Additional Considerations
The Case	15 minutes	Student packet	Could be read in class or as a homework assignment prior to class
Investigate	15 minutes	Student packet, container filled with at least 1 in. of water, additional water, strips of black paper, 1 bottle of clear nail polish, paper towels, indirectly vented chemical splash safety goggles, nonlatex apron, nitrile gloves	Small-group or individual activity
Apply and Analyze	15 minutes	Student packet, internet access	Whole-class, small-group, or individual activity
Design Challenge	30 minutes	Student packet	Small-group activity

Teacher Background Information

This case is intentionally written in such a way that a thorough knowledge of the nature of light and physics is not required. This structure makes it possible to present the activity to students of different ages and in different course contexts (e.g., biology or physics). The content in this case is well suited to complement a unit delving into a more quantitative treatment of the characteristics of electromagnetic radiation, especially the wavelength of visible light and constructive and destructive interference.

Vocabulary

- cellulose
- iridescence/iridescent
- nano- (nanoscale, nanometer)

- pigment
- thin film
- wavelength

Teacher Answer Key

Recognize, Recall, and Reflect

1. **What was the surprising observation Silvia Vignolini made about the *Pollia condensata* berry that prompted her to learn more about it?**

    Vignolini noticed that the Pollia condensata *berries shone with an intense blue color and was surprised to discover the berries had been picked over forty years before. This made her wonder how the berry was able to keep its intense iridescent color that lasted many years.*

2. **In your own words, what causes the blue iridescence of the *Pollia condensata* berry?**

    The specific size and shape of the helix (spiral staircase) nanoscale surface features cause the berry to only reflect blue light.

3. **What property of the berry has to change in order to show other colors such as red, purple, or green?**

    The size of the spacing of the cellulose layers in the helix (spiral staircase) would have to change size in order to reflect different colors of light.

Questions for Reflection

1. **What colors do you observe in the film-covered strip of paper? What happens when you move the strip around and look at it from different angles?**

    Student answers may vary, but typically students report seeing all colors visible in the rainbow.

2. **You applied a thin layer of clear nail polish to black paper. In your own words, explain how this process has created the colors you are observing.**

    The very thin layer of nail polish causes a thin film interference with the light that hits the surface of the polish. The film is a few hundred nanometers thick. The polish is thicker in some places than others and reflects different wavelengths of light (i.e., different colors) in different locations.

3. **If you wanted to create a thin film that was all the same color (green, for example), what would need to be different about your thin film?**

    The film on the paper would need to be same thickness throughout. (If the film is different thicknesses in different locations, you will see many colors.) The film would need to be the specific thickness required to reflect green light, in order to see that color.

Apply and Analyze

1. **When did the *Microraptor* live?**

   The Microraptor *lived about 120 million years ago.*

2. **What clues do scientists look for in a fossil when trying to determine the color of an extinct animal?**

   Imprints of feathers in the fossil can indicate the shape of melanin structures, called melanosomes. Scientists measure the distance between the melanosomes. This distance, measured in nanometers, tells them which wavelength/color of light is reinforced when white light shines on the nanostructure. (For example, green light has a wavelength of about 500 nm, so melanosomes that are about 500 nm across will reflect green light. Thinner melanosomes will give off purple light; thicker melanosomes will give off red light.) Scientists may try to take many samples from different locations on the fossil to determine if the feathers were all one color. By comparing the melanosome imprint in the fossil to the melanosomes in modern birds, scientists might find similarities between their sample and a modern bird, which may also help them in their analysis.

3. **Can scientists use this technique to determine the color of all extinct animals from fossils?**

   This technique only works if the fossil is extremely well-preserved and if the fossilized creature exhibited structural color. If the creature's color came from pigments in its skin or feathers, those molecules would have degraded and will not have been preserved in the fossil.

Reflect

1. **What technologies might need to be developed to create or manufacture this design?**

   Student answers will vary.

2. **What are any constraints or drawbacks you can foresee with implementing this design?**

   Student answers will vary.

3. **Would there be any environmental or human health concerns to using structural color in this way?**

   Student answers will vary.

Assessment

The Design Challenge can be assessed using the rubric in the appendix (p. 377). You might also facilitate an activity to assess student understanding of the observable characteristics of structural color in natural specimens. Provide students with several samples (different types of feathers, butterfly wings, beetle shells, etc.) to examine, and ask students to determine if the color they observe is due to pigment or due to nanostructure. Ask students to provide evidence for their reasoning.

Extensions

- As an extension activity, you can provide students with different natural specimens that get their colors from either pigments or from structural color. Have students explore and compare the various specimens. Peacock feathers exhibit structural color and are found at most craft stores. There are also many videos of wildlife exhibiting colors that shift when viewed from different angles. For instance, check out this video from PBS showing the mating dance of the male Lawes's parotia: *www.pbs.org/video/nature-lawess-parotia*.

- For follow-up activities that involve thin film interference, have your students blow bubbles and describe why they observe a rainbow of colors moving across the surface of the bubble.

Resources and References

Ball, P. 2012. Nature's color tricks: Understanding seven clever tactics animals use to create dazzling hues may lead to sophisticated new technologies. *Scientific American* 306 (5): 74–79.

Bloch, L. et al. "How do we know what color dinosaurs were?" TED-Ed video. *https://ed.ted.com/lessons/how-do-we-know-what-color-dinosaurs-were-len-bloch*.

Fox, D. 2016. What sparked the cambrian explosion? *Nature. www.nature.com/news/what-sparked-the-cambrian-explosion-1.19379*.

Luiggi, C. 2013. Color from structure. *The Scientist. www.the-scientist.com/?articles.view/articleNo/34200/title/Color-from-Structure*.

PBS Nature. 2011. "Lawes's parotia." PBS video. *www.pbs.org/video/nature-lawess-parotia*.

Parker, A. R. 2000. 515 million years of structural colour. *Journal of Optics A: Pure and Applied Optics* 2 (6): R15–R18.

FROM SLUDGE TO NOBILITY

Artificial Dye

A Case Study Using the Discovery Engineering Process

Introduction

How did you spend your last spring break? In 1856, 18-year-old William Henry Perkin discovered the first synthetic dye from chemicals originating from coal tar while on vacation from London's Royal College of Chemistry. Perkin's discovery was accidental; he made it as he was attempting to synthesize quinine, a drug that fights malaria.

Lesson Objectives

By the end of this case study, you will be able to

- Describe how artificial dye was discovered.

- Explain how indicators work.

- Analyze the pH of household materials using red cabbage.

- Design a new application for indicators that helps to solve a problem.

The Case

Read the following summary of the history of artificial dye.

For centuries, the color purple was synonymous with power, wealth, and royalty due to the cost associated with producing the dye. Before 1856, dyes were produced from natural materials, such as plants and insects. Only a few colors were available, including orange, yellow, green, and purple. Purple dye was the most expensive, as thousands of marine mollusks were needed to produce a small amount of dye. (The dye was made from secretions by these mollusks.)

In the mid-1800s, scientists at London's Royal College of Chemistry became interested in using coal tar to create new products. Coal tar is a black or dark-brown by-product that is created when processing coal into other fuel products. Coal tars can be thick liquids or semisolids. There are many uses for coal tar. For instance, waterproof fabric was first invented by applying a rubbery solution of coal tar to a layer of cloth.

Perkin wanted to use coal tar to make synthetic quinine. Quinine was a common treatment for malaria naturally produced from the cinchona tree in South America.

FIGURE 3.1

Wool Being Dyed

While experimenting, Perkin accidentally created a black residue. Further experiments with this black residue led to a new product that was a rich purple color. This gave Perkin an idea. He used the new purple compound to dye a piece of silk. This method of dying silk was not only cheap and easy, but the color did not fade when exposed to sunlight. Nor did it run when exposed to water.

Perkin called his discovery Tyrian purple. Dyeing factories of the time were skeptical about using this new dye. So, alongside his father and brother, Perkin began producing his own dye, which he now called mauve. Mauve became available to the lower classes due to its affordability. Today, synthetic dyes are often used to add color to clothing and other products (Figure 3.1).

Recognize, Recall, and Reflect

1. What was Perkin originally researching?

2. Why was purple dye so expensive to make?

3. What characteristics made Perkin's dye appealing to the lower class?

Investigate

This section allows you to experiment with color-changing substances. In this activity, you will explore using indicators to determine a solution's pH level—that is, its level of acidity or basicity. An indicator is a substance that undergoes an observable change (for instance, a change in color) when the conditions of its solution change. Like dyes, indicators were originally extracted from various sources, including plant pigments. In this activity, you will use the pigment flavin that is in red cabbage to test the acidity or basicity of various solutions.

Note: Today a common indicator used to tell whether a solution is an acid or base is litmus paper. (The word *litmus* comes from Norse—the language of ancient Scandinavia—and means "to color or dye.") The indicator litmus is red in acidic solutions when a pH is less than 7 and blue in basic solutions when the pH is greater than 7. If these strips are available, you will use them during the activity as well.

Materials

For each group of students:

- Red cabbage indicator liquid provided by teacher
- Small containers (glass is recommended; one container is needed for each item to test pH)
- Variety of items to test pH (fruit juice, soda, vinegar, baking soda solution, etc.)
- Dropper
- Indirectly vented chemical splash safety goggles (1 pair per student)
- Nonlatex apron (1 per student)
- Nitrile gloves (1 pair per student)
- pH indicator strips (optional)

Safety Note: Wear indirectly vented chemical splash safety goggles, a nonlatex apron, and nitrile gloves during the setup, hands-on, and takedown segments of the activity. Wash your hands with soap and water immediately after completing this activity.

Create, Innovate, and Investigate

- Begin by examining the indicator solution.
- Pour your indicator solution into the small containers.

- Using your dropper, add the liquid you want to test to one of your indicator solution cups. You may need to swirl the cup to see the solution change colors.

- When you see a color change, use Table 3.1 to determine the pH of your liquid. Then record your observations in the chart below.

- **Extension:** If you have pH indicator strips, use the strips to determine the pH of the solution. Was your red cabbage indicator accurate?

TABLE 3.1

Liquid pH Levels

pH	Color
2	Red (very acidic)
4	Purple
6	Violet
7	Indicator color (neutral)
8	Blue
10	Blue-green
12	Greenish-yellow (very basic)

Indicator Observation Chart

Item	Color	pH

Questions for Reflection

1. What did you observe about the original color of the red cabbage indicator?

2. What happened when you added your solutions to the indicator?

Apply and Analyze

Acids and bases are used when dyeing fabrics. Acidic dyes are water-soluble (i.e., they can be dissolved in water). They are used on silk, wool, and nylon fabrics. Vinegar, which contains acetic acid, is used to bond the dye to the fibers. Basic dyes are water-soluble, as well. They are applied to acrylic fibers and are also used to color paper. Basic dyes are usually aided by acetic acid to help the dye adhere to the fabric. Visit the listed websites to learn more about fabric dye, and then answer the questions that follow.

- *www.quilthistory.com/dye.htm*
- *www.theguardian.com/sustainable-business/sustainable-fashion-blog/2015/mar/31/*
  natural-dyes-v-synthetic-which-is-more-sustainable

1. How was the art of natural dyeing lost?
2. What are some of the cost-versus-benefit concerns of using synthetic dyes instead of natural dyes?

Design Challenge

Engineering is the application of scientific understanding through creativity, imagination, problem solving, and the designing and building of new materials to address and solve problems in the real world. You will be asked to take the science you have learned in this case and design a process or product to address a real-world issue of your choosing.

Engineers use the engineering design process as steps to address real-world problems (see Figure 3.2). You will now use this process as you come up with a new application for an indicator. In this case, you are asking the question (Step 1) of how indicators can be used for new purposes. Drawing on your creativity,

FIGURE 3.2

The Engineering Design Process

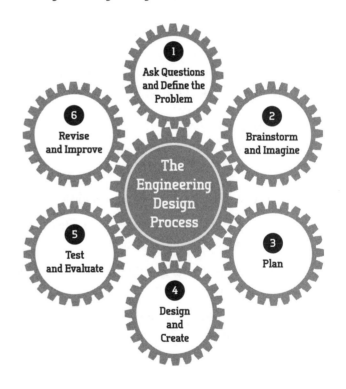

you will imagine (Step 2) a specific new application for an indicator. Afterward, you will create a plan (Step 3) for this new product. Next, you will create a sketch and/or model of your product (Step 4). Then, you will work with your classmates to think about how you would test (Step 5) and refine (Step 6) your product.

1. Ask Questions

Based on your research, consider a new problem that may be addressed or a product that could be created by using an indicator. What are some applications where you would need a tool that visually indicates a change that would otherwise go unnoticed?

2. Brainstorm and Imagine

Brainstorm a useful product that could employ indicator technology. (For example, indicators are used to test swimming pools and soil, but these have to be tested by someone manually. Developing an indicator that could automatically alert you if a pool's pH was too high or too low would be helpful for hotels instead of relying on an employee to continually test the pool.) Also, keep in mind that indicators do not always have to test for pH; they could also test for temperature changes, pressure, or other characteristics.

3. Create a Plan

Create a plan for your new product. Consider: (1) What is the purpose of your product? (2) What are the benefits of the product? (3) What are the limitations of the product? Use the Product Planning Graphic Organizer (p. 50) to help you.

4. Design and Create

Consider the following questions and considerations for your indicator product and its design.

- How would incorporating an indicator make this product better?

- Are there any limitations or drawbacks to using indicators? If so, how would you overcome them?

- What technologies might need to be developed to create or manufacture this design?

- What are any constraints or drawbacks you can foresee with implementing this design?

- Would there be any safety concerns regarding your indicator-based product?

Now create a sketch of your indicator design. Make sure your design incorporates the research and exploration you've done.

5. Test and Evaluate

Working with your classmates, come up with a way to test your design to see its effectiveness.

6. Revise and Improve

Give your plans to one of your classmates for review. Listen to his or her feedback on your design and take some time to revise and make improvements. What are some ways you can use the input and refine your design?

Reflect

1. What technologies might need to be developed to create or manufacture this design?

2. What are any constraints or drawbacks you can foresee with implementing this design?

3. Would there be any environmental or human health concerns about using indicators in this way?

Product Planning Graphic Organizer

Proposed Product Idea

Pros (Benefits)	Cons (Limitations)

TEACHER NOTES

FROM SLUDGE TO NOBILITY

ARTIFICIAL DYE

A Case Study Using the Discovery Engineering Process

Lesson Overview

In this lesson, students learn how artificial dye was accidentally discovered and subsequently used in new applications. Then, they examine indicators as a way to test pH levels. They also learn how acids and bases are used in dying fabrics.

Lesson Objectives

By the end of this case study, students will be able to

- Describe how artificial dye was discovered.

- Explain how indicators work.

- Analyze the pH of household materials using red cabbage.

- Design a new application for indicators that helps to solve a problem.

The Case Study Approach

This lesson uses a case study approach. Explaining the purpose of case studies will encourage your students to relate to the material and engage with the problem. At the heart of each case study in this book is a true story, one that describes how someone in his or her everyday life or during a routine workday made an observation or did a simple experiment that led to a new insight or discovery. Case studies are designed to get students actively engaged in the process of problem solving. The narrative of the case supplies authentic details that place the student in the role of the inventor and provide scaffolds for critical thinking and deep reflection. A case is more than a paragraph to read or a story to analyze but rather a way of framing problems, synthesizing what is known, and thinking creatively about new applications and solutions. In this lesson, students consider how a specific clothing dye was discovered and work together to think about new applications for dyes and indicators to solve real-life problems.

Use of the Case

Due to the nature of these case studies, teachers may elect to use any section of each case for their instructional needs. The sections are sequenced in order (scaffolded) so students think more deeply about the science involved in the case and develop an understanding of engineering in the context of science.

Curriculum Connections

Lesson Integration

You could use this case as a way to integrate engineering design into a lesson on acids, bases, and indicators. This lesson can also be used with Chapter 2, "Harnessing the Color of Nature: Structural Color and Iridescence" (p. 25).

Related Next Generation Science Standards

PERFORMANCE EXPECTATIONS

- HS-PS2-6. Communicate scientific and technical information about why the molecular-level structure is important in the functioning of designed materials.

- HS-ETS1-2. Design a solution to a complex real-world problem by breaking it down into smaller, more manageable problems that can be solved through engineering.

- HS-ETS1-3. Evaluate a solution to a complex real-world problem based on prioritized criteria and trade-offs that account for a range of constraints, including cost, safety, reliability, and aesthetics, as well as possible social, cultural, and environmental impacts.

SCIENCE AND ENGINEERING PRACTICES

- Analyzing and Interpreting Data
- Constructing Explanations and Designing Solutions

CROSSCUTTING CONCEPT

- Structure and Function

Related National Academy of Engineering Grand Challenge

- Engineer the Tools of Scientific Discovery

Lesson Preparation

You will need to make copies of the entire student section for the class. Students will need internet access at various points in the lesson. Alternatively, you can project videos or print and distribute copies of online content for the class. Prior to the Investigate activity, you will need to grate a red cabbage into a bowl or container. Next, pour just enough boiling water over the cabbage to cover it. Steep the mixture until the liquid is room temperature, stirring occasionally. Using a strainer, pour the contents into another large bowl. You should now have liquid that is a reddish purple. Students will need a portion of the indicator liquid for each item they test. Litmus strips are optional but can be purchased from a variety of sources, including Carolina Biological Supplies and Educational Innovations. Look at the Teaching Organizer (Table 3.2, p. 54) for suggestions on how to organize the lesson.

Materials

For teacher:

- 2 large bowls
- Red cabbage
- Grater
- Boiling water
- Strainer

For each group of students:

- Red cabbage indicator liquid provided by teacher
- Small containers (one for each item to test pH)
- Variety of items to test pH (fruit juice, soda, vinegar, baking soda solution, etc.)
- Dropper
- Indirectly vented chemical splash safety goggles (1 pair per student)
- Nonlatex apron (1 per student)
- Nitrile gloves (1 pair per student)
- pH indicator strips (optional)

Safety Note for Students: Wear indirectly vented chemical splash safety goggles, a nonlatex apron, and nitrile gloves during the setup, hands-on, and takedown segments of the activity. Wash your hands with soap and water immediately after completing this activity.

Time Needed

Up to 60 minutes

TABLE 3.2

Teaching Organizer

Section	Time Suggested	Materials Needed	Additional Considerations
The Case	5 minutes	Student packet	Could be read in class or as a homework assignment prior to class
Investigate	10–15 minutes	For teacher: 2 large bowls, red cabbage, grater, boiling water, strainer For students: class packet, indicator liquid, small containers (one for each item to test pH), variety of items to test pH (fruit juice, soda, vinegar, baking soda solution, etc.), dropper, indirectly vented chemical splash safety goggles, nitrile gloves, nonlatex apron, pH indicator strips (optional)	Recommended as a small-group activity; teacher will need to prep red cabbage before class.
Apply and Analyze	10 minutes	Student packet, internet access	Small-group or individual activity
Design Challenge	30 minutes	Student packet	Small-group activity

Teacher Background Information

Students often have questions about pH, acids, bases, and indicators. Khan Academy has helpful resources on acids and bases: *www.khanacademy.org/science/chemistry/acids-and-bases-topic*. You can also use the following video from Flinn Scientific to help explain how to test acid and base indicators: *www.youtube.com/watch?v=7zCqWUJBWAA*.

Vocabulary

- acid
- base
- flavin
- indicator
- pH
- solution

Teacher Answer Key

Recognize, Recall, and Reflect

1. **What was Perkin originally researching?**

   Medication to treat malaria

2. **Why was purple dye so expensive to make?**

   Thousands of marine mollusks were needed to make a small amount of dye.

3. **What characteristics made Perkin's dye appealing to the lower class?**

   It was cheap, did not fade in sunlight, and did not run when exposed to water.

Questions for Reflection

1. **What did you observe about the original color of the red cabbage indicator?**

   The color will change based on the presence of an acid or a base.

2. **What happened when you added your solutions to the indicator?**

   The pigment will turn red in an acidic environment (pH less than 7) and bluish-green in a basic environment (pH greater than 7).

Apply and Analyze

1. **How was the art of natural dyeing lost?**

   After Perkin's discovery of synthetic dye, many dyers abandoned natural dyeing methods and adopted new methods that involved synthetic dye.

2. **What are some of the cost-versus-benefit concerns of using synthetic dyes instead of natural dyes?**

   Natural dyes are generally more eco-friendly but harder to produce. Synthetic dyes are easier to create and last longer, but factories that make and use these dyes produce wastewater, a significant source of pollution.

Reflect

1. **What technologies might need to be developed to create or manufacture this design?**

   Answers may vary depending on student designs.

2. **What are any constraints or drawbacks you can foresee with implementing this design?**

Answers could include the elimination of jobs due to automatic indicators.

3. **Would there be any environmental or human health concerns to using indicators in this way?**

Answers may vary depending on student designs.

Assessment

The Design Challenge can be assessed using the rubric in the appendix (p. 377). Students could also be asked to write a report on the benefits and limitations of using litmus paper and plant indicators.

Extensions

This lesson can be followed with lessons about chemical reactions.

Resources and References

ATSDR. 2018. Relevance to public health. CDC. *www.atsdr.cdc.gov/toxprofiles/tp85-c2.pdf*.

Chemical Heritage Foundation. 2016. William Henry Perkin. *www.chemheritage.org/historical-profile/william-henry-perkin*.

Chhabra, E. 2015. Natural dyes v synthetic: Which one is more sustainable? TheGuardian.com. *www.theguardian.com/sustainable-business/sustainable-fashion-blog/2015/mar/31/natural-dyes-v-synthetic-which-is-more-sustainable*.

ColorantsHistory.org. 2015. William H. Perkin founder of dyestuff industry. *http://colorantshistory.org/PerkinBiography.html* (accessed January 11, 2018).

Dharma Trading Co. Did you know ... how acid dyes work. *www.dharmatrading.com/home/did-you-know-how-acid-dye-works.html* (accessed January 11, 2018).

Driessen, K. Quilt history: The earliest dyes. *www.quilthistory.com/dye.htm* (accessed February 5, 2019).

Flinn Scientific. 2015. "How to test acid-base indicators." YouTube video. *www.youtube.com/watch?v=7zCqWUJBWAA*.

Gilbert, M. 2017. History of fabric dyes. Our Pastimes. *https://ourpastimes.com/the-history-of-fabric-dyes-12215773.html*.

Khan Academy. Acids and bases. *www.khanacademy.org/science/chemistry/acids-and-bases-topic* (accessed February 5, 2019).

Podhajny, R. 2002. History, shellfish, royalty, and the color purple. Paper, Film and Foil Converter. *http://pffc-online.com/ar/1348-paper-history-shellfish-royalty*.

Science Buddies Staff. 2017. Cabbage chemistry. *www.sciencebuddies.org/science-fair-projects/project-ideas/Chem_p013/chemistry/make-cabbage-pH-indicator*.

Victorian Web. 2014. Sir William Henry Perkin and the coal-tar colours. *www.victorianweb.org/science/perkin.html*.

BY THE TEETH OF YOUR SKIN

Shark Skin and Bacteria

4

A Case Study Using the Discovery Engineering Process

Introduction

Sharks, like most fish, have scales on their skin. However, shark scales are special. They're bony and contain substances called dentine and enameloid, which is similar to the enamel on human teeth. (See Figure 4.1, p. 58.) These unique scales are called dermal denticles, which in layman's terms literally means "skin teeth." Dermal denticles cause sharks to feel smooth from head to tail but rough in reverse. A shark's scales serve many purposes. They help the shark move faster through water and allow it to be stealthy by reducing the noise the shark makes as it swims. Dermal denticles also prevent bacteria and other organisms from growing on sharks. Today, researchers are designing products that mimic shark skin and its useful properties. Read more on shark scales here: *elasmo-research.org/education/white_shark/scales.htm.*

Lesson Objectives

By the end of this case study, you will be able to

- Describe the structure of shark scales.

- Explain why a rough surface can prevent bacterial growth.

- Create a product or process that utilizes the same properties as shark skin for reducing drag or bacteria growth.

The Case

Read the following summary of how new products were made to mimic shark skin. After reading the summary, answer the questions that follow.

Naval ships are large and require a lot of maintenance to remain ready for duty. One issue they face is the growth of organisms such as algae, barnacles, or tunicates on their bottom sides. These organisms cause drag, which reduces the ship's speed and increases the amount of fuel needed to move the ship. It also costs a lot of time and money to move naval ships to areas where they can be dry-docked, or taken out of the water, and cleaned.

FIGURE 4.1

Close-Up of Shark Scales

In 2002, Dr. Anthony Brennan, a University of Florida materials science and engineering professor, visited the U.S. naval base at Pearl Harbor in Oahu, Hawaii, to participate in research sponsored by the U.S. Navy. While he was there, the U.S. Office of Naval Research asked Brennan to find new strategies to cut costs associated with dry-docking and drag caused by algae growth.

In order to find a solution, Brennan looked to the ocean. After noticing that submarines resembled slow-moving whales due to the buildup of algae and barnacles on their surface, he wondered if there were any large marine animals that do not have such organisms growing on their skin. The only one he could think of was the shark. Brennan took an impression of shark skin to look at under his microscope. He observed that the dermal denticles on the skin created a pattern of dips and ridges. The ratio of height to width of these structures creates a course surface where it is very difficult for algae and other organisms to thrive. Inspired by his findings, Brennan created a synthetic surface that mimicked shark skin and named it Sharklet (see Figure 4.2). In its first test, the product was able to decrease green algae settlement by 85%.

Since it worked so well in discouraging algae growth, Brennan wanted to see if Sharklet could prevent the growth of other microorganisms. The product turned out to be very successful in keeping away bacteria. This may be because growing up and down the Sharklet's pattern of ridges takes more energy than the bacteria can afford to use, so it must colonize elsewhere. Sharklet is now being used in many medical products, such as catheters, to prevent bacterial infections.

Illustration of Magnified Sharklet Material

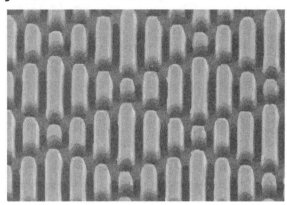

Recognize, Recall, and Reflect

1. What was the problem Dr. Brennan was trying to solve?

2. What sparked Brennan's idea to use shark skin as a model?

3. By how much did Sharklet decrease green algae growth?

4. How does Sharklet keep bacteria from growing?

Investigate

Healthcare-associated infections (HAIs) are major infections patients get in healthcare settings while being treated for other issues. One source of HAIs includes bacteria growth on hospital equipment. Items such as catheters or drainage tubes are especially prone to growing biofilms, which are responsible for about 80% of all bacterial infections and are very difficult to treat. Two examples of common bacteria found in healthcare settings are

*   *Pseudomonas aeruginosa,* a bacterium found in the environment that can cause sepsis (an infection of the bloodstream), pneumonia, and infections surrounding surgical sites; and

*   *Staphylococcus aureus,* a bacterium normally found in a person's nose, which can cause sepsis, pneumonia, bone infections, or infections of the heart valves.

To learn more about strategies for reducing HAIs, researchers designed an investigation to compare the number of bacteria cells growing on different types of surfaces. Two types of bacteria were grown on 25 plates, each with one of three surfaces: a smooth surface, shark skin, and a ridged surface that mimicked shark skin. The researchers then used an electron microscope to count the number of cells growing on each plate. The average number of cells on each type of surface was recorded as a percentage of the number of cells on the smooth surface. (See Figure 4.3.)

FIGURE 4.3

Bacterial Cell Growth by Type of Surface

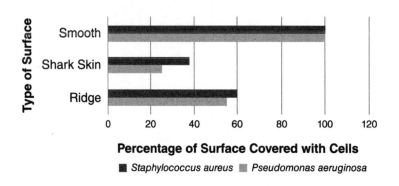

Percentage of Surface Covered with Cells

■ *Staphylococcus aureus* ▨ *Pseudomonas aeruginosa*

Source: Based on information from Sakamoto, A. et al. 2014. Antibacterial effects of protruding and recessed shark skin micropatterned surfaces of polyacrylate plate with a shallow groove. *FEMS Microbiology Letters* 361 (1): 10–16.

Questions for Reflection

Using the graph, answer the following questions:

1. Which surface had the lowest percentage of bacterial cells?

2. How much more effective was the ridge surface than the smooth surface at preventing bacterial growth for *Staphylococcus aureus*?

Apply and Analyze

While catheters are one of the first hospital instruments being developed with Sharklet to prevent infections, there are many other opportunities to prevent infections by creating products from better materials. Every day, an average of 4% of hospital patients have an HAI, according to the Center for Disease Control. Table 4.1 lists common HAIs and the methods by which they are frequently transferred.

TABLE 4.1

Common Healthcare-Associated Infections

Infections	Methods of Transfer
Acinetobacter baumannii	This bacterium is commonly found in soil or water but infections rarely happen outside of healthcare facilities. The infections can be caused by contact with infected people or surfaces. They can also be caused by contaminated breathing machines, catheters, or other equipment. Open wounds are at risk for infections, as well. For more information visit *www.cdc.gov/hai/organisms/acinetobacter.html*.
Klebsiella pneumoniae	This bacterium is normally found in human intestines. When a person comes into contact with the bacteria outside of his or her body, it can cause pneumonia, sepsis, meningitis, and infections surrounding surgical sites. To cause an infection, the bacteria must enter through the respiratory tract or blood stream. Patients with catheters or breathing machines or patients on long-term antibiotics are most at risk for an infection caused by *Klebsiella pneumoniae*. For more information visit *www.cdc.gov/hai/organisms/klebsiella/klebsiella.html*.
Mycobacterium abscessus	This bacterium is found in water, soil, and dust. It can contaminate medicines and medical devices, and it is often transferred when injections are given without properly cleaning the injection site first. This bacterium typically infects the skin, the soft tissues under the skin, and the lungs in people with weakened immune systems or other respiratory diseases. Infections can also occur when an open wound is exposed to contaminated dirt in the outdoors. For more information visit *www.cdc.gov/hai/organisms/mycobacterium.html*.
Norovirus	Norovirus causes an infection in the intestines, which leads to severe vomiting and diarrhea. Dehydration is a major side effect of the illness. Norovirus is caused by a virus, meaning it cannot be treated with antibiotics. Hospitals are at risk for the spread of norovirus due to the frequency of person-to-person contact. Eating contaminated food or drinks or touching contaminated surfaces can also lead to infection. For more information visit *www.cdc.gov/hai/organisms/norovirus.html*.
Pseudomonas aeruginosa	Infection from this bacterium is most commonly seen in people who have weakened immune systems. Hospital patients with catheters, breathing machines, or surgical wounds are particularly at risk. Some infections can also come from contact with infected water, leading to ear infections or rashes. For more information visit *www.cdc.gov/hai/organisms/pseudomonas.html*.
Staphylococcus aureus	Staph is spread through contact with a person infected with the bacteria, inhaling infected droplets from an infected person's sneezes, or through the use of an object contaminated by the bacteria. For more information visit *www.cdc.gov/hai/organisms/staph.html*.

Source: www.cdc.gov/hai/organisms/organisms.html.

Using the information in Table 4.1 (p. 61), answer the following questions:

1. Based on what you've read about HAIs, what appears to be the most common mode of infection?

2. Using materials that mimic shark skin might reduce the risk of infection for which of these HAIs? How would the material reduce infection risk in these cases?

3. Other than catheters made from materials that mimic shark skin, what other types of hospital equipment might be altered to reduce infections?

Design Challenge

Engineering is the application of scientific understanding through creativity, imagination, problem solving, and the designing and building of new materials to address and solve problems in the real world. You will be asked to take the science you have learned in this case and design a process or product to address a real-world issue of your choosing.

FIGURE 4.4

The Engineering Design Process

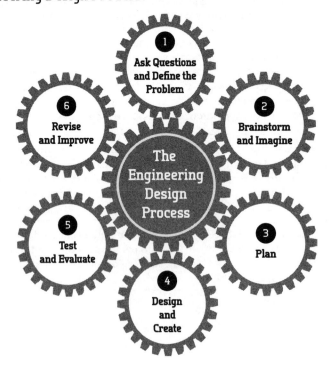

Engineers use the engineering design process as steps to address real-world problems (see Figure 4.4). You will now use this process as you come up with a new way to use a shark skin–inspired material. In this case, you are asking the question (Step 1) of how you can design a new use for shark skin–inspired material. Drawing on your creativity, you will then brainstorm (Step 2) a new product that uses a shark skin–inspired material to solve a problem. Afterward, you will create a plan (Step 3) for this new product. Next, you will create a sketch and/or model of your product (Step 4). Then, you will think about how you would test (Step 5) and refine (Step 6) your product.

1. Ask Questions

Based on your research, consider a new problem that may be addressed or a product that could be created by using a shark skin–inspired material. What are some applications where you would need a material that is resistant to the growth of bacteria and other organisms?

2. Brainstorm and Imagine

Brainstorm a specific product that uses shark skin–inspired material to solve a problem. (For instance, perhaps a shark skin–inspired material could be developed to create bacteria-resistant cooking surfaces, cell phone covers that decrease the risk of infection, or bacteria-resistant bathroom door handles.)

3. Create a Plan

Now, you and your group will create a business plan for developing a new product inspired by shark skin. Using the Business Plan Form (p. 65), outline the process needed to develop your new product. First, you will need to explain your rationale, or reasons, for developing the product. What problem are you trying to solve and why? Next, describe the product and how it will solve your problem. Then list the materials needed to develop your product. Once you have developed your product, how will you test it to make sure it works? Finally, list any questions, concerns, or issues regarding your product that need to be addressed. Think about how you will respond to them.

4. Design and Create

You and your teammates will collect the materials needed to make your product from your homes. You can use sandpaper provided by your teacher or other rough materials to mimic shark skin. Once you have all your materials, use a sheet of paper to sketch out your design. Then build a model of your design based on this sketch. Consider the following questions and considerations for your shark skin–inspired product and its design.

1. How would decreasing the growth of bacteria or other organisms make this product better?

2. How would you test whether the product was able to decrease the amount of bacterial growth?

3. Are there any limitations to using this material as part of your product? If so, how would you overcome them?

4. What technologies might need to be developed to create or manufacture this design?

5. What are any constraints or drawbacks you can foresee with implementing this design?

6. Would there be any safety concerns regarding your shark skin–inspired product?

Safety Note: Wear indirectly vented chemical splash safety goggles, a nonlatex apron, and nitrile gloves during the setup, hands-on, and takedown segments of the activity.

5. Test and Evaluate

Once you have built your model, you will present it to your classmates, who will act as potential investors. Your goal is to encourage them to fund the development of your product. Create a presentation in any format you choose to showcase your design. Make sure you discuss the important ideas you developed in your business plan. The "investors" will use the New Product Evaluation Form (p. 66) to evaluate your product.

Once you have presented your product and received feedback, you will play the role of the potential investor for your classmates. Make sure to give constructive feedback so that your peers can improve their designs, as well.

6. Revise and Improve

Consider your classmates' feedback on your design and take some time to revise and make improvements. What are some ways you can use their input to refine your design?

Reflect

1. What technologies might need to be developed to create or manufacture this design?

2. What are any constraints or drawbacks you can foresee with implementing this design?

3. Would there be any environmental or human health concerns about using shark skin–inspired material in this way?

Business Plan Form

Name: _____

RATIONALE	What is the problem and why must we solve it?

PROCESS	Describe the product and how it will solve the issue.

MATERIALS	What is needed to create the product?

TESTING	How will you test the product to make sure it works?

OTHER ISSUES	What other issues, questions, or concerns do you have that need to be addressed?

New Product Evaluation Form

Name: _____

1	What is the new product?

2	What problem is it trying to solve?

3	Do the presenters create a compelling argument for using this product to solve their problem?

4	Do they give an appropriate plan for evaluating the product?

5	Are there any questions that the presentation did not answer? If so, list them below.

6	Should your company fund this product? Why or why not?

7	What recommendations would you make to improve this product?

NATIONAL SCIENCE TEACHERS ASSOCIATION

TEACHER NOTES

BY THE TEETH OF YOUR SKIN

SHARK SKIN AND BACTERIA

A Case Study Using the Discovery Engineering Process

Lesson Overview

In this lesson, students will learn about a product developed to mimic shark skin. This product is used in the medical field to prevent the growth of bacteria on equipment and in shipbuilding to prevent the growth of algae and other microorganisms on ship bottoms. Students will learn how this material was developed and brainstorm ways to solve an everyday problem with it.

Lesson Objectives

By the end of this case study, students will be able to

- Describe the structure of shark scales.

- Explain why a rough surface can prevent bacterial growth.

- Create a product or process that utilizes the same properties as shark skin for reducing drag or bacteria growth.

The Case Study Approach

This lesson uses a case study approach. Explaining the purpose of case studies will encourage your students to relate to the material and engage with the problem. At the heart of each case study in this book is a true story, one that describes how someone in his or her everyday life or during a routine workday made an observation or did a simple experiment that led to a new insight or discovery. Case studies are designed to get students actively engaged in the process of problem solving. The narrative of the case supplies authentic details that place the student in the role of the inventor and provide scaffolds for critical thinking and deep reflection. A case is more than a paragraph to read or a story to analyze but rather a way of framing problems, synthesizing what is known, and thinking creatively about new applications and solutions. In this lesson, students consider how shark skin–inspired material was discovered and work together to think about new applications for this type of material.

Use of the Case

Due to the nature of these case studies, teachers may elect to use any section of each case for their instructional needs. The sections are sequenced in order (scaffolded) so students think more deeply about the science involved in the case and develop an understanding of engineering in the context of science.

Curriculum Connections

Lesson Integration

You could use this case as a way to integrate engineering into lessons on how the environment influences the growth of organisms. It may be useful to review information related to bacteria. This lesson may also be used with the physical sciences to discuss how synthetic materials are made from natural resources. Information on biomimicry will support such lessons as well.

Related Next Generation Science Standards
PERFORMANCE EXPECTATIONS

- MS-PS1-3. Gather and make sense of information to describe that synthetic materials come from natural resources and impact society.

- MS-LS1-5. Construct a scientific explanation based on evidence for how environmental and genetic factors influence the growth of organisms.

- MS-ETS1-1. Define the criteria and constraints of a design problem with sufficient precision to ensure a successful solution, taking into account relevant scientific principles and potential impacts on people and the natural environment that may limit possible solutions.

SCIENCE AND ENGINEERING PRACTICES

- Asking Questions and Defining Problems
- Developing and Using Models

Lesson Preparation

You will need to make copies of the entire student section for the class. Students will need internet access at various points in the lesson. Alternatively, you can project videos or print and distribute copies of online content for the class. You will need sandpaper, which you can easily obtain from a local hardware store. Also, students will need to collect materials from home for their design challenge. No list of materials for the challenge is supplied in order to allow students to think creatively about what materials they may want to use. Look at the Teaching Organizer (Table 4.2) for suggestions on how to organize the lesson.

Materials

- Sandpaper

- Building materials for model

Safety Note for Students: Wear indirectly vented chemical splash safety goggles, a nonlatex apron, and nitrile gloves during the setup, hands-on, and takedown segments of the activity.

Time Needed

2–3 class periods

TABLE 4.2

Teaching Organizer

Section	Time Suggested	Materials Needed	Additional Considerations
The Case	5 minutes	Student packet, internet access	Could be read in class or as a homework assignment prior to class
Investigate	10 minutes	Student packet	Could be read in class or as a homework assignment prior to class
Apply and Analyze	10 minutes	Student packet, internet access	Small-group or individual activity
Design Challenge	Up to 2 class periods	Student packet, sandpaper, coloring supplies, building materials for models	Small-group activity

Teacher Background Information

Sharks have lived on planet Earth since before the dinosaurs. They are well adapted to their environment. One of their unique adaptations is their skin. Shark skin is covered in scales called dermal denticles. These specialized scales allow sharks to be more hydrodynamic, and they prevent growth of other organisms on their skin.

Different species of sharks have uniquely shaped dermal denticles, which creates a variety of skin textures. By studying dermal denticles under the microscope, researchers were able to identify the properties of shark skin. They then developed materials that mimicked shark skin and incorporated these materials into new products to make them resistant to the growth of organisms. These new products include medical equipment, such as catheters, that can help prevent hospital infections. They have also been used in shipbuilding to reduce drag and time out of water for repairs. New products, such as antibacterial cell phone cases, are still in the works.

Vocabulary

- bacteria
- biomimicry
- dermal denticle

Teacher Answer Key

Recognize, Recall, and Reflect

1. **What was the problem Dr. Brennan was trying to solve?**

   He was trying to decrease the amount of algae, barnacles, and tunicates that grow on the bottom of boats.

2. **What sparked Brennan's idea to use shark skin as a model?**

   He realized that sharks were the only large ocean creatures that do not have algae growing on them.

3. **By how much did Sharklet decrease green algae growth?**

   85%

4. **How does Sharklet keep bacteria from growing?**

   Sharklet's ridged surface keeps away bacteria, probably because it takes too much energy for microorganisms to grow or spread across the ridges.

Questions for Reflection

1. **Which surface had the lowest percentage of bacterial cells?**

   Shark skin

2. How much more effective was the ridge surface than the smooth surface at preventing bacterial growth for *Staphylococcus aureus*?

40%

Apply and Analyze

1. Based on what you've read about HAIs, what appears to be the most common mode of infection?

Person-to-person contact or coming into contact with contaminated surfaces

2. Using materials that mimic shark skin might reduce the risk of infection for which of these HAIs? How would the material reduce infection risk in these cases?

Answers may vary; however, any bacteria that spreads through contact could be reduced by using shark skin–inspired materials to decrease bacteria growth.

3. Other than catheters made from materials that mimic shark skin, what other types of hospital equipment might be altered to reduce infections?

Answers may vary. They may include things like doorknobs, countertops, or surgical gloves.

Reflect

1. What technologies might need to be developed to create or manufacture this design?

Answers will vary depending on the design.

2. What are any constraints or drawbacks you can foresee with implementing this design?

Answers will vary depending on the design.

3. Would there be any environmental or human health concerns from using shark skin–inspired material in this way?

Answers will vary depending on the design.

Assessment

Students will peer-assess each other's projects using the New Product Evaluation Form (p. 66). By evaluating one another and using peer feedback to improve their design, they will have a better understanding of the process new products go through.

Extensions

Have students research other examples of products designed after ocean creatures. Search terms should include *biomimicry* (that is, the development of products that mimic properties found in nature).

Resources and References

Landhuis, E. 2014. Repelling germs with "sharkskin." Science News for Students. *https:// student.societyforscience.org/article/repelling-germs-%E2%80%98sharkskin%E2%80%99.*

ReefQuest Center for Shark Research. Skin of the teeth. *www.elasmo-research.org/education/ white_shark/scales.htm* (accessed September 27, 2018).

Sharklet. Inspired by nature: The discovery of Sharklet. *www.sharklet.com/our-technology/ sharklet-discovery* (accessed September 27, 2018).

5

A STICKY SITUATION

Gecko Feet Adhesives

A Case Study Using the Discovery Engineering Process

Introduction

Geckos are a type of lizard found all around the world. (See Figure 5.1.) Scientists have been studying the feet of geckos since it was discovered that the reptiles could hang by one toe from a piece of glass. The secret to how they were able to do so was a mystery until recently. Scientists have discovered how to make an adhesive that mimics the properties of a gecko foot. This has allowed them to develop products that can be used in a variety of situations where traditional adhesives may not work or would damage objects.

FIGURE 5.1

Gecko

Lesson Objectives

By the end of this case study, you will be able to

- Describe how van der Waals forces work.

- Analyze the tack (stickiness) of different adhesives.

- Design a new use for gecko-inspired adhesives.

The Case

Read the following summary of an adhesive developed using the same principles as those used by gecko feet.

Since the time of the Romans, scientists have been studying the feet of geckos. In the early 2000s, researchers learned that gecko feet have microscopic hairs called setae (see Figure 5.2). These hairs attach to objects through the use of van der Waals forces. Van der Waals forces are relatively weak electric forces that involve the attraction of neutral atoms or molecules. These forces are distance-dependent and weaker than covalent, ionic bonds, and metallic bonds. Covalent bonds are formed when electrons are shared between atoms. Ionic bonds are formed when one atom gives up one or more electrons to another atom. Metallic bonds hold metals together and are formed by the attraction of fixed, positively charged metallic atoms and mobile electrons that move around the entire structure.

FIGURE 5.2

Up-Close Photo of a Gecko Foot

Even though scientists had discovered that geckos use van der Waals forces to climb and hang from surfaces, they struggled with how to use that information to make new forms of adhesives. Researchers at the University of Massachusetts Amherst were studying the problem from two angles. A biologist named Duncan Irschick was investigating how geckos can cling to various objects. At the same time, a polymer scientist named Alfred Crosby was trying to make adhesives using the principle of van der Waals forces. The two began working together to develop a solution.

Dr. Irschick discovered that the setae on gecko feet had stiff tendons connected to them under the skin. Dr. Crosby was able to use this information to develop a new adhesive made up of stiff fibers attached to a flexible polymer. The fibers drape over surfaces, which allows them to bond through van der Waals forces. A slight pull in a different direction releases the adhesive. Their new product, named Geckskin, allows objects that weigh as much as 700 pounds to be attached vertically to smooth surfaces like glass with as little as a note card–size piece of adhesive.

Recognize, Recall, and Reflect

1. What structural feature of a gecko's feet allows the reptile to stick to walls?

2. How are van der Waals forces different from other forces?

3. How did the two researchers from the University of Massachusetts Amherst work together to create a gecko-inspired adhesive?

Investigate

While gecko-inspired adhesives are not currently available to the public, many other adhesives are on the market for consumers. In this activity, you will test the tack, or stickiness, of these different products to explore their adhesive properties. Record your observations on the Adhesive Testing Chart (p. 77).

Materials

For each group of students:

- 1 ruler
- 2 chairs
- 1 quart-size plastic bag
- Pennies or small weights
- 2 adhesive bandages
- Masking tape
- Cellophane tape
- Duct tape
- First-aid tape
- Electrical tape
- Packing tape
- Scissors

Safety Note: Do not allow your feet or other parts of the body to remain under the plastic bag while testing the tack of the different adhesives.

Create, Innovate, and Investigate

- Begin by examining each of the adhesives. Look at their shapes, colors, and textures. What properties do you notice?

- Next, cut a small piece from each of the different types of tape.

- Touch each of the adhesives. What do you observe about the tack, or stickiness, of each one? Predict which materials are stickiest based on your initial observations, rating them on a scale of 1 to 7 on your chart (with 1 being the stickiest and 7 being the least sticky). Once you have made your predictions, you will test each material.

- Set two chairs back-to-back, about 6 inches apart.

- Balance a ruler on the backs of the two chairs so it forms a bridge.

- Cut a 4-inch strip of each adhesive.

- You will test each adhesive, one at a time. Attach the first adhesive to the top edge of a plastic bag, using one inch at the strip's end. (*Note:* Use a new strip of tape, *not* the one you examined earlier.)

- Then attach the bag to the ruler using one inch of the other end of the adhesive. The bag should remain open.

- Gently place the pennies or weights in the bag, one at a time, until the bag falls from the adhesive. Write the number of pennies or weights it took to pull the bag from the adhesive on your chart.

- Repeat the activity with each of the other adhesives.

- Once you have collected your data, compare your results to your initial prediction as to which adhesive was strongest.

Adhesive Testing Chart

Name: _____

Material	Initial Observations	Stickiness Ranking	Stickiness Prediction	Number of Pennies
Adhesive bandage				
Masking tape				
Cellophane tape				
Duct tape				
First-aid tape				
Electrical tape				
Packing tape				

Questions for Reflection

1. What did you observe about the different types of adhesives?

2. Which adhesive was strongest? Were you surprised? Why or why not?

3. Which adhesive was weakest? Were you surprised? Why or why not?

4. How would you decide which type of adhesive to use in the future?

5. Why is it important to know what type of adhesive to use?

Apply and Analyze

The concept of gecko-inspired materials is quickly being adopted for new areas of research and commercial applications. Watch this video on how the gecko-inspired adhesive created by Irschick and Crosby works: *www.youtube.com/watch?v=9ZJYbcG0Ts0*. Next, visit the links below to explore a set of articles and videos created by *Science* magazine, which show how researchers are developing new products that mimic gecko feet. After you're done reading the articles and watching the videos on these web pages, answer the questions that follow.

- Watch astronauts play catch with a gecko-inspired gripper: *www.sciencemag.org/news/2017/06/watch-astronauts-play-catch-gecko-inspired-gripper*.

- See how gecko-inspired adhesives allow people to climb walls: *www.sciencemag.org/news/2014/11/gecko-inspired-adhesives-allow-people-climb-walls*.

1. How many times can someone use Geckskin before it loses its adhesive properties?

2. Why are you unable to use traditional adhesives in space?

3. The inventors using gecko-inspired adhesives believe the technology can be used to create robots that assist astronauts in space or hold heavy objects on vertical surfaces. What are some other new applications that you can imagine for gecko-inspired adhesives in your classroom?

Design Challenge

Engineering is the application of scientific understanding through creativity, imagination, problem solving, and the designing and building of new materials to address and solve problems in the real world. You will be asked to take the science you have learned in this case and design a process or product to address a real-world issue of your choosing.

Engineers use the engineering design process as steps to address a real-world problem (see Figure 5.3). You will now use this process as you come up with a new way to use a gecko-inspired adhesive. In this case, you are asking the question (Step 1) of how you can design a new use for a gecko-inspired adhesive. Drawing on your creativity, you will then brainstorm (Step 2) a new product that uses a gecko-inspired adhesive to solve a problem. Afterward, you will create a plan (Step 3) for this new product. Next, you will create a sketch and/or model of your product (Step 4). Then, you will work with your classmates to think about how you would test (Step 5) and refine (Step 6) your product.

FIGURE 5.3

The Engineering Design Process

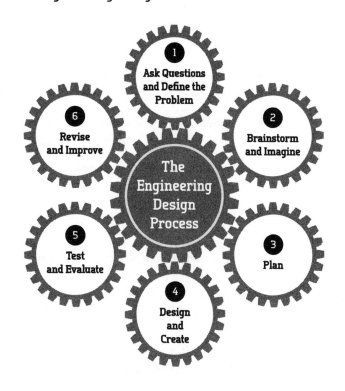

1. Ask Questions

Based upon your previous research, consider a new problem that may be addressed or product that could be created by using materials inspired by gecko feet. Think about applications where you need a material that can stick, be easily removed, and not damage the material to which it is sticking. What problem could be solved with a material like this?

2. Brainstorm and Imagine

Brainstorm a specific application for a gecko-inspired adhesive that could help solve a problem. (For example, perhaps such an adhesive could be used to hang wall art so that the art doesn't damage the wall.)

3. Create a Plan

Create a plan for your product. Consider: (1) What is the purpose of the product? (2) What are benefits to using this product? (3) What are the limitations of using this product? (For example, if you were to create a gecko-inspired adhesive that could be used to hang wall art, you'd have to come up with the pros and cons of the product.) Use the Product Planning Graphic Organizer to help you.

4. Design and Create

Consider the following questions and considerations for your product and its design.

- How would incorporating a gecko-inspired adhesive into your design make the product better?

- Are there any limitations or drawbacks to using a gecko-inspired adhesive? If so, how would you overcome them?

- What technologies might need to be developed to create or manufacture this design?

- What are any constraints or drawbacks you can foresee with implementing this design?

- Would there be any safety concerns regarding your product?

Now, create a sketch of your product design. Make sure your design incorporates your previous research and exploration.

5. Test and Evaluate

Working with your classmates, come up with a way to test your design to see its effectiveness.

6. Revise and Improve

Give your plans to one of your classmates for review. Listen to his or her feedback on your design. What are some ways you can use the input to refine your design? Take some time to revise and make improvements.

Reflect

1. What technologies might need to be developed to create or manufacture this design?

2. What are any constraints or drawbacks you can foresee with implementing this design?

3. Would there be any environmental or human health concerns about this design?

Product Planning Graphic Organizer

Proposed Product Idea	
Pros (Benefits)	**Cons (Limitations)**

A STICKY SITUATION

GECKO FEET ADHESIVES

A Case Study Using the Discovery Engineering Process

Lesson Overview

In this lesson, students explore gecko foot–inspired adhesives. Geckos are able to hang by one toe from a smooth surface and researchers have finally unlocked the reason why. They are using this information to develop new adhesive products.

Lesson Objectives

By the end of this case study, students will be able to

- Describe how van der Waals forces work.

- Analyze the tack (stickiness) of different adhesives.

- Design a new use for gecko-inspired adhesives.

The Case Study Approach

This lesson uses a case study approach. Explaining the purpose of case studies will encourage your students to relate to the material and engage with the problem. At the heart of each case study in this book is a true story, one that describes how someone in his or her everyday life or during a routine workday made an observation or did a simple experiment that led to a new insight or discovery. Case studies are designed to get students actively engaged in the process of problem solving. The narrative of the case supplies authentic details that place the student in the role of the inventor and provide scaffolds for critical thinking and deep reflection. A case is more than a paragraph to read or a story to analyze but rather a way of framing problems, synthesizing what is known, and thinking creatively about new applications and solutions. In this lesson, students consider how adhesives inspired by gecko feet were discovered and work together to think about new applications for such adhesives to solve real-life problems.

Use of the Case

Due to the nature of these case studies, teachers may elect to use any section of each case for their instructional needs. The sections are sequenced in order (scaffolded) so students think more deeply about the science involved in the case and develop an understanding of engineering in the context of science.

Curriculum Connections

Lesson Integration

You could use this case as a way to integrate engineering into a lesson on atomic structure or bonds. It may be useful to review the properties of atomic forces.

Related Next Generation Science Standards

PERFORMANCE EXPECTATIONS

- MS-PS1-3. Gather and make sense of information to describe that synthetic materials come from natural resources and impact society.

- MS-ETS1-1. Define the criteria and constraints of a design problem with sufficient precision to ensure a successful solution, taking into account relevant scientific principles and potential impacts on people and the natural environment that may limit possible solutions.

- HS-PS2-6. Communicate scientific and technical information about why the molecular-level structure is important in the functioning of designed materials.

- HS-ETS1-3. Evaluate a solution to a complex real-world problem based on prioritized criteria and trade-offs that account for a range of constraints, including cost, safety, reliability, and aesthetics, as well as possible social, cultural, and environmental impacts.

SCIENCE AND ENGINEERING PRACTICES

- Analyzing and Interpreting Data
- Engaging in Argument From Evidence
- Constructing Explanations and Designing Solutions

CROSSCUTTING CONCEPT

- Structure and Function

Related National Academy of Engineering Grand Challenge

- Engineer the Tools of Scientific Discovery

Lesson Preparation

You will need to make copies of the entire student section for the class. Students will need internet access at various points in the lesson. Alternatively, you can project videos or print and distribute copies of online content for the class. Look at the Teaching Organizer (Table 5.1) for suggestions on how to organize the lesson.

Materials

For each group of students:

- 1 ruler
- 2 chairs
- 1 quart-size plastic bag
- Pennies or small weights
- 2 adhesive bandages
- Masking tape
- Cellophane tape
- Duct tape
- First-aid tape
- Electrical tape
- Packing tape
- Scissors

Safety Note for Students: Do not allow your feet or other parts of the body to remain under the plastic bag while testing the tack of the different adhesives.

Time Needed

55 minutes

TABLE 5.1

Teaching Organizer

Section	Time Suggested	Materials Needed	Additional Considerations
The Case	5 minutes	Student packet	Could be read in class or as a homework assignment prior to class
Investigate	10 minutes	Student packet, 1 ruler, 2 chairs, 1 quart-size plastic bag, pennies or small weights, 2 adhesive bandages, masking tape, cellophane tape, duct tape, first-aid tape, electrical tape, packing tape, scissors	Small-group activity. During the investigation, students may choose to add additional types of adhesives. They may also want to investigate what effect heat, cold temperatures, or water have on the adhesives.
Apply and Analyze	10 minutes	Student packet, internet access	Small-group or individual activity
Design Challenge	30 minutes	Student packet	Small-group activity

Teacher Background Information

It may be helpful to review the properties of atomic forces. Students often also have questions about bonding. These resources from the National Science Teachers Association may be of use:

- Book chapter on chemical bonds
  http://common.nsta.org/resource/?id=10.2505/9780873552738.9

- SciGuide on atomic structure and chemical bonding
  http://common.nsta.org/resource/?id=10.2505/5/SG-07

- E-book on quantitative evaluation
  http://common.nsta.org/resource/?id=10.2505/PKEB237X

Vocabulary

- adhesive
- covalent
- ionic

- setae
- tack
- van der Waals forces

Teacher Answer Key

Recognize, Recall, and Reflect

1. **What structural feature of a gecko's feet helps the reptile to stick to walls?**

    Gecko feet have tiny hairs called setae attached to tendons under the skin that help them to stick to walls.

2. **How are van der Waals forces different from other forces?**

    Van der Waals forces—electric forces that involve the attraction of neutral atoms or molecules—are relatively weaker than other forces.

3. **How did the two researchers from the University of Massachusetts Amherst work together to create a gecko-inspired adhesive?**

    Dr. Irschick discovered that the setae on gecko feet had stiff tendons connected to them under the skin. Dr. Crosby was able to use this information to develop a new adhesive made up of stiff fibers attached to a flexible polymer.

Questions for Reflection

1. **What did you observe about the different types of adhesives?**

    Answers will vary.

2. **Which adhesive was strongest? Were you surprised? Why or why not?**

    Answers will vary.

3. **Which adhesive was weakest? Were you surprised? Why or why not?**

    Answers will vary.

4. **How would you decide which type of adhesive to use in the future?**

    Answers will vary.

5. **Why is it important to know what type of adhesive to use?**

    Answers will vary but may include that you sometimes need stronger adhesives and sometimes weaker ones depending on what objects you're trying to stick together.

Apply and Analyze

1. **How many times can someone use Geckskin before it loses its adhesive properties?**

    You can use it over and over again.

2. **Why are you unable to use traditional adhesives in space?**

    Suction won't work, because most of the universe is a vacuum, and you can't use suction in a vacuum. Other traditional adhesives that use sticky chemicals are degraded by the extreme temperatures of Earth's orbit.

3. **The inventors using gecko-inspired adhesives believe the technology can be used to create robots that assist astronauts in space or hold heavy objects on vertical surfaces. What are some other new applications that you can imagine for gecko-inspired adhesives in your classroom?**

    Examples for uses may include holding a television on the wall that can be moved as the classroom is rearranged or installing movable dry erase boards.

Reflect

1. **What technologies might need to be developed to create or manufacture this design?**

    Answers will vary depending on the students' designs.

2. **What are any constraints or drawbacks you can foresee with implementing this design?**

    Answers will vary depending on the students' designs.

3. **Would there be any environmental or human health concerns about this design?**

    Answers will vary depending on the students' designs.

Assessment

The Design Challenge can be assessed using the rubric in the appendix (p. 377).

Extensions

This lesson can be combined with the case in Chapter 10, "A Sticky Discovery: The Invention of Post-It Notes" (p. 153).

Resources and References

Beiser, A. 2003. *Concepts of modern physics*. 6th ed. New Delhi: Tata McGraw-Hill Publishing Company Limited.

Green, I., K. Hughes, J. Whitmire, and A. Campbell. 2013. Teaching the scientific method using adhesives. *www.juliantrubin.com/encyclopedia/chemistry/adhesive_experiments.html*.

Jones, M. G., A. Taylor, and M. Falvo. 2009. *Extreme science: From nano to galactic*. Arlington, VA: NSTA Press.

Keeley, P., F. Eberle, and J. Tugel. 2007. Chemical bonds. In *Uncovering student ideas in science, vol. 2: 25 more formative assessment probes*, 71–75. Arlington, VA: NSTA Press.

National Science Teachers Association (NSTA). SciGuide: Atomic structure and chemical bonding. *http://common.nsta.org/resource/?id=10.2505/5/SG-07*.

Shouse, B. 2002. How geckos stick on der Waals. *Science. www.sciencemag.org/news/2002/08/how-geckos-stick-der-waals*.

University of Massachusetts Amherst. Geckskin. *https://geckskin.umass.edu/#science* (accessed January 12, 2018).

Wessel, L. 2017. Watch astronauts play catch with a gecko-inspired gripper. *Science. www.sciencemag.org/news/2017/06/watch-astronauts-play-catch-gecko-inspired-gripper*.

You, J. 2014. Gecko-inspired adhesives allow people to climb walls. *Science. www.sciencemag.org/news/2014/11/gecko-inspired-adhesives-allow-people-climb-walls*.

IT'S HIP TO BE SQUARE

Seahorse Tails

A Case Study Using the Discovery
Engineering Process

Introduction

In nature, tails play a variety of roles. For instance, kangaroos use their tails to maintain balance. Tails can also be used for communication, such as with the male peacock. Various animals hold onto things with their tails, as seen with certain monkeys, opossums, and seahorses. Seahorses are known for having prehensile tails that are able to grab and hold onto objects. These fish sport armored bodies and range in size from less than an inch to more than 8 inches in length. They use their tube-shaped mouths to slurp up organisms as they float by. Male seahorses have a unique quality: They are the only male animals in the world that give birth. Most of a seahorse's life is spent holding onto underwater vegetation and hard structures like coral to keep from being swept away by currents (see Figure 6.1). Recently, researchers noticed that seahorse tails were square in cross section, not round like most other animal tails. This has led to a variety of research projects.

FIGURE 6.1

A Seahorse Gripping With Its Tail

Lesson Objectives

By the end of this case study, you will be able to

- Describe the structure of a seahorse tail.

- Build a structure inspired by a seahorse tail.

- Create a product or process that utilizes the structure of a seahorse tail.

The Case

Read the following summary of a study on seahorse tails.

Seahorses are one of the slowest fish in the ocean. Most fish use their caudal fin, or tail, to swim through the water. Seahorses use their dorsal fin, or back fin, to propel themselves forward and their tail to move up and down or to hold onto objects. Researchers from Clemson University recently designed a study modeling seahorse tails.

While most tails in nature are rounded like a cylinder, seahorse tails have a square cross section like a square prism. The square shape of their tail helps them wrap around objects in order to stay stationary in the water. The square shape may also play a role in protecting seahorses and their relatives, the pipefish, from predation. As slow swimmers, seahorses and pipefish are prone to being eaten by predators. Pipefish are particularly at risk of being eaten by birds. Scientists hypothesize that the structure and armor found in the tails of pipefish and seahorses may be flexible enough to prevent these body parts from being crushed by bird beaks.

In order to examine the structure of a seahorse tail, Clemson University researchers used a 3-D printer to create models of both seahorse tails and other, rounded tails. They wanted to understand the seahorse tail's grasping ability and its armored properties.

For the seahorse tails, the researchers built working models that used square prisms; for the round tails, the researchers built models that used cylindrical prisms. (See Figure 6.2.) The researchers wanted to compare the two shapes for strength and flexibility. Once they made the models, they twisted, turned, bent, and banged on them with hammers. As with a real seahorse tail, the square model consisted of smaller square segments. Each of these square segments was made up of four overlapping, L-shaped plates connected at various points by gliding joints. These moveable joints allowed the plates to slide past each other when pressure was applied, making the tail stronger and preventing damage from being crushed (see Figure 6.3). The cylindrical prisms deformed under stress, which would damage a live animal. The researchers concluded that the square tail was more flexible and stronger than the round tail, and they hoped to develop new tools and robots based on their research. You can download a video of the researchers testing the square prism model here: *www.sciencemag.org/content/suppl/2015/07/01/349.6243.aaa6683.DC1/aaa6683s2.mov.*

FIGURE 6.2

Models of Different Tail Types

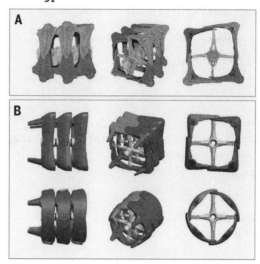

Computer models of square prisms (representing square tails) and circular prisms (representing round tails)

Source: From Porter, M. et al. 2015. Why the seahorse tail is square. *Science* 349 (6243): aaa6683. Reprinted with permission from AAAS.

FIGURE 6.3

Seahorse Tail Models

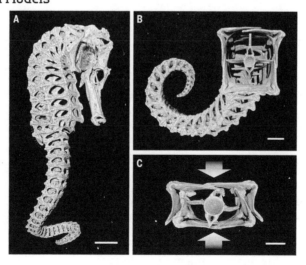

Computer-generated images of (A) a seahorse's skeleton, (B) the cross section of its tail, and (C) a crushed tail segment

Source: From Porter, M. et al. 2015. Why the seahorse tail is square. *Science* 349 (6243): aaa6683. Reprinted with permission from AAAS.

Recognize, Recall, and Reflect

1. What makes seahorse tails different from other species?

2. Why did the Clemson University scientists want to study seahorse tails?

3. What tool did they use to develop models for their research?

4. Which model tail was strongest?

Investigate

The researchers used 3-D printers to create different tail shapes. How might they have created the shapes without a 3-D printer? Let's find out. Each group will need the materials listed below.

Materials

For each group of students:

- Building materials such as balsa wood, straws, paper, etc.

- Glue, tape, rubber bands

- Scissors or scalpels for cutting

- Weights

- A structure to lay tail models on, such as a pipe

- Aluminum foil

- Wire, pipe cleaners, or wax-covered string

- Safety goggles (1 pair per student)

Safety Note: Wear safety glasses or goggles during the setup, hands-on, and take-down segments of the activity. Use caution when using sharp tools as they can cut or puncture skin. Wash your hands with soap and water immediately after completing this activity.

Create, Innovate, and Investigate

- Download this video of seahorse tail models being bent and twisted: *www. sciencemag.org/content/suppl/2015/07/01/349.6243.aaa6683.DC1/aaa6683s1.mov.*

1. What was the final alignment for each tail?

2. What would be the impact of these alignments?

3. How can you build a model to test different tail structures?

- Using the materials supplied by your teacher, build a structure that can bend but be strong enough not to crush. Consider the structure of the seahorse tail as you build it.

- Groups will be able to test and refine their designs until they feel confident that they are bendable and strong.

- Each group will then test its structure by bending it over an object selected by the teacher. Once the structure is set up, weights will be placed on top until the structure cracks. Your goal is to create a design and build a prototype that will support the most weight.

Questions for Reflection

1. What structure supported the most weight?

2. How was the winning structure different from the other structures?

Apply and Analyze

Seahorse tails may be the inspiration for several new products. Read the following summary of the article "Flexible Armor: Mysterious Seahorse Astounds Scientists" (ABC News 2013) to find out more.

Researchers at the University of California, San Diego, have studied different types of animals to see how they protect themselves. Their goal is to solve engineering problems by examining solutions found in nature. This process is called biomimetics, or the mimicry of life. The researchers specifically examined different types of armor and defenses, including fish scales, turtle shells, and deer antlers. They were hoping to develop something that could grab objects while withstanding pressure from the environment. They eventually settled on the seahorse as a source of inspiration.

Like the Clemson University researchers, the researchers from UC San Diego wanted to see how seahorses did under pressure. So, they subjected dead seahorses to crushing forces, finding that a seahorse could be compressed to half its size without permanent damage. This surprised the researchers as they expected the bony plates that make up the seahorse's armor to break. The researchers determined that the seahorse's ability to withstand pressure had to do with its bony plates. Animal bones are often mostly made of minerals, which makes them brittle. Cow bones, for instance, are more than 65% mineral. Seahorse plates are only 40% mineral, which gives them more flexibility. This means that if a bird tries to eat a seahorse but does not compress the fish by more than 50% of its original width, then the seahorse could possibly escape and survive. Using what they learned from their seahorse study, the researchers planned to make a gripping device that could operate in difficult environments with various stressors.

1.   What are some of the ways seahorse tail–inspired technology could be utilized?

2.   Are there potential negative impacts to using a design based off of seahorse tails?

Design Challenge

Engineering is the application of scientific understanding through creativity, imagination, problem solving, and the designing and building of new materials to address and solve problems in the real world. You will be asked to take the science you have learned in this case and design a process or product to address a real-world issue of your choosing.

Engineers use the engineering design process as steps to address a real-world problem (see Figure 6.4). You will now use this process as you come up with a new application for a product that mimics a seahorse's tail. In this case, you are asking the question (Step 1) of how you can design a new use for a seahorse tail–inspired product. Drawing on your creativity, you will then brainstorm (Step 2) a new product inspired by a seahorse tail to solve a problem. Afterward, you will create a plan (Step 3) for this new product. Next, you will create a sketch and/or model of your product (Step 4). Then, you will work with your classmates to think about how you would test (Step 5) and refine (Step 6) your product.

FIGURE 6.4

The Engineering Design Process

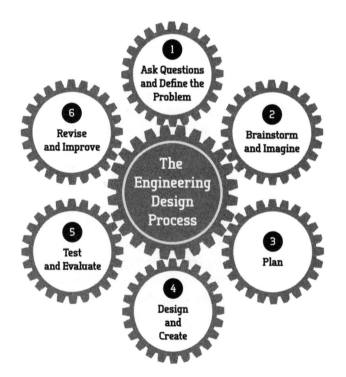

1. Ask Questions

Based on your previous research, consider a new problem that may be addressed or a product that could be created by using a material that can bend, flex, and be crushed. What are some applications where you would need a material that has the flexibility to bend or be crushed but can remain strong? What problem could be solved with this material? Think big (e.g., large robots); think small (e.g., tiny surgical instruments).

2. Brainstorm and Imagine

Use what you have learned to brainstorm a specific seahorse tail–inspired product—that is, a product made of a material that is strong and bendable. This product should solve some sort of problem. (For example: (1) Perhaps you could use this material to make a tiny surgical instrument that is strong enough to perform an operation, yet flexible enough to enter a person's body through a smaller incision to reduce scarring. (2) Robotic arms that can snake into small spaces yet support weight might be useful for rescuing people from collapsed buildings.)

3. Create a Plan

Create a plan for your product. Consider: (1) What is the purpose of the product? (2) What are benefits to using this product? (3) What are the limitations of using this product? Use the Product Planning Graphic Organizer (p. 97) to help you.

4. Design and Create

Sketch your design onto a piece of paper. Draw your model from more than one angle so you can visualize how it will look in three dimensions. You may want to sketch more than one version. Next, you will build a model of your new product. Your teacher will provide you with the necessary building materials. First you will use a piece of wire or pipe cleaners to build the frame for your product. Continue to mold it until it fits your sketch. Once you have built the frame for your product, wrap it in aluminum foil to make your product more three-dimensional.

5. Test and Evaluate

Once you have built your model, you will present it to a group of three classmates. Explain your design and the purpose for your product. As a group, discuss ways to test your design. You will not test your model, but it is important to discuss how you might test the actual product.

6. Revise and Improve

Listen to your group's feedback on your design and take some time to revise and make improvements. What are some ways you can use their input to refine your design? You may want to use more foil or wire to adjust your final product.

Reflect

1. What technologies might need to be developed to create or manufacture this design?

2. What are any constraints or drawbacks you can foresee with implementing this design?

3. Would there be any environmental or human health concerns about using a seahorse-inspired material in this way?

Product Planning Graphic Organizer

Proposed Product Idea	
Pros (Benefits)	**Cons (Limitations)**

IT'S HIP TO BE SQUARE

SEAHORSE TAILS

A Case Study Using the Discovery Engineering Process

Lesson Overview

In this lesson, students learn about the unique properties of a seahorse's tail. They learn how the shape and structure of the tail is being used to develop new products such as gripping devices that could operate in difficult environments. The students will build and test their own flexible yet strong structures.

Lesson Objectives

By the end of this case study, students will be able to

- Describe the structure of a seahorse tail.

- Build a structure inspired by a seahorse tail.

- Create a product or process that utilizes the structure of a seahorse tail.

The Case Study Approach

This lesson uses a case study approach. Explaining the purpose of case studies will encourage your students to relate to the material and engage with the problem. At the heart of each case study in this book is a true story, one that describes how someone in his or her everyday life or during a routine workday made an observation or did a simple experiment that led to a new insight or discovery. Case studies are designed to get students actively engaged in the process of problem solving. The narrative of the case supplies authentic details that place the student in the role of the inventor and provide scaffolds for critical thinking and deep reflection. A case is more than a paragraph to read or a story to analyze but rather a way of framing problems, synthesizing what is known, and thinking creatively about new applications and solutions. In this lesson, students consider how researchers studied the seahorse tail structure and work together to think of new ways in which the structure can be used to solve real-life problems.

Use of the Case

Due to the nature of these case studies, teachers may elect to use any section of each case for their instructional needs. The sections are sequenced in order (scaffolded) so students think more deeply about the science involved in the case and develop an understanding of engineering in the context of science.

Curriculum Connections

Lesson Integration

You could use this case as a way to integrate engineering into lessons on physics concepts such as force, compression, torsion, gravity, and motion. This activity can also be used to discuss adaptations, form, and function.

Related Next Generation Science Standards

PERFORMANCE EXPECTATIONS

- MS-ETS1-1. Define the criteria and constraints of a design problem with sufficient precision to ensure a successful solution, taking into account relevant scientific principles and potential impacts on people and the natural environment that may limit possible solutions.

- MS-ETS1-2. Evaluate competing design solutions using a systematic process to determine how well they meet the criteria and constraints of the problem.

- HS-ETS1-3. Evaluate a solution to a complex real-world problem based on prioritized criteria and trade-offs that account for a range of constraints, including cost, safety, reliability, and aesthetics as well as possible social, cultural, and environmental impacts.

SCIENCE AND ENGINEERING PRACTICES

- Analyzing and Interpreting Data

- Engaging in Argument From Evidence

- Constructing Explanations and Designing Solutions

- Developing and Using Models

CROSSCUTTING CONCEPTS

- Cause and Effect

- Structure and Function

Lesson Preparation

You will need to make copies of the entire student section for the class. Students will need internet access at various points in the lesson. Alternatively, you can project videos or print and distribute copies of online content for the class. Before beginning this lesson, you will need to collect materials for building seahorse-tail models and new products based off of seahorse tails. To make the lesson more inquiry based, you should select a variety of materials for the students to choose from. Some suggested items are listed below. If needed, review information on bridge-building contests to give you ideas to help your students. Look at the Teaching Organizer (Table 6.1) for suggestions on how to organize the lesson.

Materials

For each group of students during the Investigate section and for each individual student during the Design Challenge:

- Building materials such as balsa wood, straws, paper, etc.

- Glue, tape, rubber bands

- Scissors or scalpels for cutting

- Weights (only needed for the Investigate section)

- A structure to lay tail models on, such as a pipe (only needed for the Investigate section)

- Aluminum foil

- Wire, pipe cleaners, or wax-covered string

- Safety goggles (1 pair per student)

Safety Note for Students: Wear safety glasses or goggles during the setup, hands-on, and takedown segments of the activity. Use caution when using sharp tools as they can cut or puncture skin. Wash your hands with soap and water immediately after completing this activity.

Time Needed

3 class periods

TABLE 6.1

Teaching Organizer

Section	Time Suggested	Materials Needed	Additional Considerations
The Case	5 minutes	Student packet, internet access	Could be read in class or as a homework assignment prior to class
Investigate	10 minutes	Student packet; internet access; building materials such as balsa wood, straws, paper, etc.; glue, tape, rubber bands; scissors or scalpels for cutting; weights; a structure to lay tail models on, such as a pipe; aluminum foil; wire, pipe cleaners, or wax-covered string; safety goggles	Small-group activity
Apply and Analyze	10 minutes	Student packet	Small-group or individual activity
Design Challenge	2 class periods	Student packet; internet access; building materials such as balsa wood, straws, paper, etc.; glue, tape, rubber bands; scissors or scalpels for cutting; aluminum foil; wire, pipe cleaners, or wax-covered string; safety goggles	Students individually create models, then share them in groups.

Teacher Background Information

Seahorses are fish. However, they have several characteristics that set them apart from most other fish. Like a chameleon, they can see in two directions at once. They have a fused mouth similar to a straw. Rather than scales, they have armored bony plates like an armadillo. Seahorses also have a prehensile tail. But unlike most other animals with this feature, their tail is square in cross section rather than round. This allows them to have a better grasp on objects. Seahorses are one of the slowest fish in the ocean and grasp onto seagrass or other structures to keep from being swept away by the currents. Researchers are currently studying the structure of seahorse tails to learn what makes them so strong and flexible. Made up of interlocking, moveable plates, the seahorse tail can be compressed to about half its width before sustaining permanent damage. This may be used to develop new types of body armor. The articulation of their tails is also being studied to develop new types of robotics. This new line of research may have great implications for the future.

Vocabulary

- articulated
- biomimetics
- cross section
- prehensile
- square prism

Teacher Answer Key

Recognize, Recall, and Reflect

1. **What makes seahorse tails different from other species?**

   Unlike most other animals, which have round cross sections, seahorses have square cross sections.

2. **Why did the Clemson University scientists want to study seahorse tails?**

   To examine the structure of the tails and find out whether they were stronger than round tails

3. **What tool did they use to develop models for their research?**

   3-D printers

4. **Which model tail was strongest?**

   The square model

Create, Innovate, and Investigate

1. **What was the final alignment for each tail model?**

   Square: Linearly aligned
   Cylindrical: Misaligned

2. **What would be the impact of these alignments?**

   Misalignment would mean damage or injury to the animal.

3. **How can you build a model to test different tail structures?**

   Answers will vary.

Questions for Reflection

1. **What structure supported the most weight?**

   Answers will vary.

2. **How was the winning structure different from the other structures?**

   Answers will vary based on the shape and design of each structure.

Apply and Analyze

1. **What are some of the ways seahorse tail–inspired technology could be utilized?**

   Answers may include that the technology can be used to create robots, surgical instruments, and more.

2. **Are there potential negative impacts to using a design based off of seahorse tails?**

   Answers will vary.

Reflect

1. **What technologies might need to be developed to create or manufacture this design?**

   Answers will vary.

2. **What are any constraints or drawbacks you can foresee with implementing this design?**

   Answers will vary.

3. **Would there be any environmental or human health concerns about using a seahorse-inspired material in this way?**

   Answers will vary.

Assessment

The Design Challenge can be assessed using the rubric in the appendix (p. 377).

Extensions

- If your school has access to 3-D printers, have students attempt to re-create the structures made by the researchers. Then have the students test the strength of their structures.

- Students can use online modeling software for their design challenges. A list of software options can be found here: *https://3dprinterchat.com/2017/01/13-best-cad-programs-for-kids.*

- Visit an aquarium to see seahorses in real life. Have the students look for other structures on other animals that might be good models for new products.

- Have advanced students measure the stress and strain exerted on their models. Your students will then be able to calculate Young's modulus.

Resources and References

ABC News. 2013. Flexible armor: Mysterious seahorse astounds scientists. *http://abcnews. go.com/Technology/flexible-armor-mysterious-seahorse-astounds-scientists/story?id=19151944.*

Porter, M. M., D. Adriaens, R. L. Hatton, M. A. Meyers, and J. McKittrick. 2015. Why the seahorse tail is square. *Science* 349 (6243): 30–31.

Sax, A. 2013. "How to build a balsa wood bridge." YouTube video. *www.youtube.com/ watch?v=nz_6NZ7jKj4.*

Segura, D. 2017. 13 best CAD programs for kids. 3D Printer Chat. *https://3dprinterchat. com/2017/01/13-best-cad-programs-for-kids.*

UC San Diego News Center. 2015. Why the seahorse's tail is square and how it could be an inspiration for robots and medical devices. UC San Diego. *http://ucsdnews.ucsd.edu/ pressrelease/why_the_seahorses_tail_is_square_and_how_it_could_be_an_inspiration.*

CORN FLAKES

Waste Not, Want Not

A Case Study Using the Discovery
Engineering Process

Introduction

Corn (also known as maize) is an important staple food all around the world. Corn
is a cereal grain—a type of grass that can
produce edible crops and is cultivated and
manufactured for many uses, including
the production of corn syrup, popcorn,
and flour. Corn has also been transformed
into popular breakfast cereals, such as Kellogg's Corn Flakes. Corn flakes are created
from toasted dough. Since Kellogg's introduced corn flakes to the food market, different versions of the cereal have been cre-

ated, such as Kellogg's Frosted Flakes (coated with sugar). Other cereal companies
have developed their own lines of breakfast foods created from corn.

While most individuals think of corn as a food source for humans and animals, it
has also been manufactured as an alternative to fossil fuels. Ethanol fuel can be created from corn. It's often mixed with gasoline to use as fuel for vehicles. However,
there is a growing concern about using corn as a biofuel. Because there is a demand
for farmers to produce biofuels, the corn-based food supply may be reduced. That

affects the price of corn-based food. Although there is disagreement about the use of corn-based products, corn is an important source of energy on Earth.

Lesson Objectives

By the end of the case study, you will be able to

- Describe the characteristics and uses of corn (source of food and energy).

- Explain the process of creating cereal as a food source.

- Analyze the benefits and limitations of corn as a food source and as an energy source for living organisms.

- Design a new application for corn-based materials that solves a current problem.

The Case

Read the following summary about the history of corn flakes. This account outlines the accidental discovery of corn flakes and how corn flakes have been transformed into one of the most popular breakfast items in the United States. Once you are finished reading, answer the questions that follow.

John Harvey Kellogg was a physician and superintendent of Battle Creek Sanitarium in Michigan during the late 1800s. (Sanitariums were institutions that combined medical facilities and spa-like wellness centers to offer treatments and therapies to individuals.) One common recommendation at Battle Creek Sanitarium was for patients to practice a vegetarian, or meat-free, diet. Kellogg worked with different types of grains and doughs to create the perfect breakfast food for the patients. He wanted to produce a grain-based "health food" that was tasty, easy to prepare, and easy to digest. Kellogg thought that cooking the grains at a very high temperature would make them easier and faster for the body to digest[1].

While experimenting with ingredients one day, Kellogg and his brother Will Keith Kellogg supposedly left out boiled wheat for too long, and it went stale. The Kellogg brothers decided to stretch out the material and bake it. When they inserted the dough into a roller (a tool that presses dough into a very thin layer), it fell apart into tiny flakes. The brothers toasted the flakes and served them to the sanitarium's patients, who enjoyed the new cereal. The Kelloggs continued to experiment with the design for the flakes until they came to resemble the product you find in grocery stores today. (Now corn flakes are made from milled corn, or corn that has been ground down into a flour; malt, or dried cereal grains; and sugar.)

---

1 Many physicians and nutritionists now agree that a slow-digesting, high-fiber cereal is a healthful breakfast option because high-fiber cereals keep people feeling full for longer periods of time and do not cause a spike in blood sugar, as more processed cereals do.

In 1906, Will Keith Kellogg decided to leave the sanitarium to begin the Battle Creek Toasted Corn Flake Company, which would later become the Kellogg Company. Corn flakes were mass marketed as a new food to U.S. consumers, and the success of corn flakes enabled the design and marketing of new cereals, such as Rice Krispies in 1928.

Recognize, Recall, and Reflect

1. What is the importance of corn? Describe a problem related to corn-based products.

2. Why was John Harvey Kellogg experimenting with cereal grains and doughs?

3. How were corn flakes invented?

Investigate

For this activity, you will imagine that you are a food scientist (an individual who uses chemistry to produce new foods or redesign existing foods).

Create, Innovate, and Investigate

You have been informed that your food manufacturing company needs a dough recipe for a new cereal. You will review the four recipes and the data collected from food testers to determine which recipe will sell the best. (See Table 7.1, pp. 108–109.) You will also need to include recommendations on how to improve the recipes.

TABLE 7.1

Food Test Results

Cereal	Ingredients	Baking Instructions	Food Tester Comments
Recipe 1: Chocolate Chip Flakes	Milled corn, salt, sugar, water, milk chocolate chips	1. Mix milled corn, water, salt, and sugar in a bowl until it is a dough. 2. Flatten dough and then break the dough into small pieces to create corn flakes. 3. Add the milk chocolate chips on top of the flakes. 4. Bake the flakes (with the chocolate chips on top) until they are crispy. 5. Let the flakes cool, then serve.	Food Tester #1: The cereal was too soggy due to the chocolate chips that melted on the flakes. Food Tester #2: The cereal was too sweet, and I could only eat a couple of bites. Food Tester #3: While the flakes tasted good, the chocolate on the flakes did not mix well with the cold milk.
Recipe 2: Strawberry and Honey Flakes	Milled corn, salt, sugar, water, honey, thinly sliced strawberries	1. Mix milled corn, water, salt, and sugar in a bowl until it is a dough. 2. Add the sliced strawberries to the dough and mix thoroughly. 3. Flatten dough and then break the dough into small pieces to create corn flakes. 4. Bake the flakes until they are crispy. 5. Remove the flakes from the oven and add a thin layer of honey on top. 6. Let the flakes cool, then serve.	Food Tester #1: The cereal was crispy and sweet. Food Tester #2: Before placing the cereal into the milk, it was crispy, and the honey was a nice addition. But once the cereal was in the milk, the strawberry slices separated from the cereal. Food Tester #3: The strawberries tasted burnt, as if they had been baked too long. Burnt strawberries do not taste good, even with the honey added.

Continued

Table 7.1 (*continued*)

Cereal	Ingredients	Baking Instructions	Food Tester Comments
Recipe 3: Spicy Flakes	Milled corn, salt, granulated sugar, water, chili pepper flakes	1. Mix milled corn, water, salt, and chili pepper flakes in a bowl until it is a dough. 2. Flatten dough and then break the dough into small pieces to create corn flakes. 3. Bake the flakes until they are crispy. 4. Remove the flakes from the oven and sprinkle a layer of sugar on top. 5. Let the flakes cool, then serve.	Food Tester #1: The sugar on the flakes was too grainy, and it made the milk very sweet. Food Tester #2: The pepper flakes were very spicy. I love spicy foods, but I wonder if our consumers will want spicy cereal in the morning. Food Tester #3: The grainy sugar on top of the flakes and the pepper did not taste good together. However, the sweet, cold milk balanced the spicy pepper in the cereal.
Recipe 4: Orange Citrus Flakes	Milled corn, salt, sugar, freshly squeezed orange juice, orange peel shavings	1. Mix milled corn, freshly squeezed orange juice, salt, and orange peel shavings in a bowl until it is a dough. 2. Flatten dough and then break the dough into small pieces to create corn flakes. 3. Bake the flakes until they are crispy. 4. Remove the flakes from the oven, let them cool, and then serve.	Food Tester #1: Cold milk and pieces of orange do not taste good together. Food Tester #2: The orange peel shavings made the cereal taste too tangy and bitter. It did not mix well with milk. Food Tester #3: The cereal with orange flavoring tastes good out of the box. It does not have the best taste when added to milk. So, I would rather eat this cereal without milk.

Questions for Reflection

Consider the recipes you just reviewed, comparing the pros and cons of each one. Then answer the questions below.

1. Which recipe do you think would sell the best in a wide market? Why?

2. Which recipe do you think would be most appealing to children? Why? Which recipe would be most appealing to adults? Why?

After reviewing the recipes and the food testers' comments, think about how you would redesign each recipe to improve taste. Write your feedback in the Design Improvement Chart on page 110.

Design Improvement Chart

Cereal Recipe	How Would You Change the Recipe to Improve the Taste of the Cereal?
Recipe 1: **Chocolate Chip Flakes**	
Recipe 2: **Strawberry and Honey Flakes**	
Recipe 3: **Spicy Flakes**	
Recipe 4: **Orange Citrus Flakes**	

Create Your Own Cereal

Follow the steps below to design your own cereal. You will have to list all of the ingredients, research and explain *why* you included those ingredients, describe how you will combine the ingredients together, and outline how the final cereal product should taste and who will eat it.

Step 1

Describe your cereal idea and who will eat it (e.g., adults, teens, or children?).

What will your cereal look and taste like? What makes it unique?

Describe your target audience. How will you market your cereal to your target audience?

Step 2

List your ingredients and explain *why* each ingredient is needed to make your cereal.

Ingredient	Why Do You Need This Ingredient?

NATIONAL SCIENCE TEACHERS ASSOCIATION

Step 3

Describe the creation and baking process for your cereal.

How will you mix and bake the ingredients together?

Step 4

Describe how you will test your new cereal.

How will you go about testing your cereal?

What types of questions would you ask the food testers when they are tasting your creation?

Apply and Analyze

Kirsten Hoskissen is a professional taste tester for a food company. She's tasked with making sure new cereal recipes taste good. In order to do her job, she must practice her observation, data collection, and communication skills while tasting foods. Read the article found in this link: *www.theguardian.com/money/2010/jun/19/ working-life-food-taster*. Then answer the questions that follow.

1. Why should food companies have taste testers? What is the importance of tasting foods before they go to market?

2. When Kirsten Hoskissen reviews cereals, what specifically does she observe and test in order to determine if the cereal is good for future sales?

3. What are potential challenges of being a food tester?

Design Challenge

Engineering is the application of scientific understanding through creativity, imagination, problem solving, and the designing and building of new materials to address and solve problems in the real world. You will be asked to take the science you have learned in this case and design a process or product to address a real-world issue of your choosing.

Engineers use the engineering design process as steps to address a real-world problem (see Figure 7.1). You will now use this process as you come up with a new application for a corn product. In this case, you are asking the question (Step 1) of how you can design a new type of corn product. Drawing on your creativity, you will then brainstorm (Step 2) a specific new corn product that could be used to solve a problem. Then, you will create a plan (Step 3) for your product. Although you will not actually produce your product, you will create a sketch and/or a model of it (Step 4). Then, you will

FIGURE 7.1

The Engineering Design Process

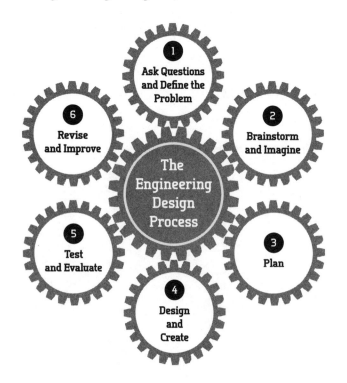

work with your classmates to think about how you would test (Step 5) and refine (Step 6) your product.

1. Ask Questions

Corn has been used to create cereals and other food products. It has also been transformed into an energy source for vehicles. Consider a new problem that may be addressed or a product that could be created using corn. (For example, could corn provide energy to power newly developed electronics? Could it be employed to produce low-cost, eco-friendly fabric for clothing?)

2. Brainstorm and Imagine

Read the following article: *www.wsj.com/articles/sneakers-made-from-corn-seat-cushions-from-soybeans-1494813781*. You can also look for other online articles on innovative uses for corn. Based on your research, brainstorm a new application for corn products. What corn-based product would be useful?

3. Create a Plan

Create a plan for your new corn product. Consider: (1) What is the purpose of this product? (2) What are benefits to using this product? (3) What are the limitations of using the product? Use the Product Planning Graphic Organizer (p. 117) to help you develop your plan.

4. Design and Create

Consider the following questions and considerations for your product and its design.

- How would incorporating the corn into your product design make it better?

- Are there any limitations or drawbacks to using corn in your product? If so, how would you overcome them?

- What technologies might need to be developed to create or manufacture this design?

- What are any constraints or drawbacks you can foresee with implementing this design?

- Would there be any safety concerns regarding your corn-based product?

Now, create a sketch of your corn-based product. Make sure your design incorporates your research and exploration you've done.

5. Test and Evaluate

Working with your classmates, come up with a way to test your design to see its effectiveness.

6. Revise and Improve

Give your plans to one of your classmates for review. Listen to his or her feedback on your design. What are some ways you can use the input to refine your design? Take some time to revise and make improvements.

Reflect

1. What technologies might need to be developed to create or manufacture this design?

2. What are any constraints or drawbacks you can foresee with implementing this design?

3. Would there be any environmental or human health concerns to using the product in this way?

Product Planning Graphic Organizer

Proposed Product Idea	
Pros (Benefits)	**Cons (Limitations)**

CORN FLAKES

WASTE NOT, WANT NOT

A Case Study Using the Discovery Engineering Process

Lesson Overview

In this lesson, students explore the history of Kellogg's Corn Flakes and the applications of corn-based products that can provide food and energy sources for human consumption. The development and application of corn flakes was an accidental discovery that has revolutionized commercial food manufacturing.

Lesson Objectives

By the end of this lesson, students will be able to

- Describe the characteristics and uses of corn (source of food and energy).

- Explain the process of creating cereal as a food source.

- Analyze the benefits and limitations of corn as a food source and as an energy source for living organisms.

- Design a new application for corn-based materials that solves a current problem.

The Case Study Approach

This lesson uses a case study approach. Explaining the purpose of case studies will encourage your students to relate to the material and engage with the problem. At the heart of each case study in this book is a true story, one that describes how someone in his or her everyday life or during a routine workday made an observation or did a simple experiment that led to a new insight or discovery. Case studies are designed to get students actively engaged in the process of problem solving. The narrative of the case supplies authentic details that place the student in the role of the inventor and provide scaffolds for critical thinking and deep reflection. A case is more than a paragraph to read or a story to analyze but rather a way of framing problems, synthesizing what is known, and thinking creatively about new applications and solutions. In this lesson, students consider how corn flakes were invented and think about new applications for corn to solve real-life problems.

Use of the Case

Due to the nature of these case studies, teachers may elect to use any section in each case for their instructional needs. The sections are sequenced in order (scaffolded) so students think more deeply about the science involved in the case and develop an understanding of engineering in the context of science.

Curriculum Connections

Lesson Integration

You could use this case as a way to integrate engineering into a lesson on chemistry involving energy transformation, physical and chemical energy properties, and manufacturing.

Related Next Generation Science Standards

PERFORMANCE EXPECTATION

- MS-PS1-3. Gather and make sense of information to describe that synthetic materials come from natural resources and impact society.

SCIENCE AND ENGINEERING PRACTICES

- Analyzing and Interpreting Data
- Engaging in Argument From Evidence
- Constructing Explanations and Designing Solutions

CROSSCUTTING CONCEPTS

- Structure and Function
- Energy and Matter

Related National Academy of Engineering Grand Challenge

- Engineer the Tools of Scientific Discovery

Lesson Preparation

You will need to make copies of the entire student section for the class. Note that the student packet is the only material needed for this lesson. Students will need internet access at various points in the lesson. Alternatively, you can project videos or print and distribute copies of online content to the class. Look at the Teaching Organizer (Table 7.2) for suggestions on how to organize the lesson.

Time Needed

Up to 130 minutes

TABLE 7.2

Teaching Organizer

Section	Time Suggested	Materials Needed	Additional Considerations
The Case	10 minutes	Student packet	Could be read in class or as homework prior to class
Investigate	20 minutes	Student packet	Whole-class activity
Create Your Own Cereal	30–40 minutes	Student packet	Small-group activity
Apply and Analyze	10–15 minutes	Student packet, internet access	Individual activity (Students can compare results with classmates.)
Design Challenge	45 minutes	Student packet, internet access	Small-group activity

Teacher Background Information

You may wish to review the concepts of energy transformation and physical and chemical energy properties. These concepts may include topics such as trophic levels and energy in an ecosystem, agriculture and farming practices and issues of sustainability, genetically modified plants for food sources, and the chemistry of energy in plants converted into food for other organisms.

Vocabulary

- cereal grain
- maize
- malt
- milled corn
- roller
- staple
- vegetarian

Teacher Answer Key

Recognize, Recall, and Reflect

1. **What is the importance of corn? Describe a problem related to corn-based products.**

   Corn is a type of plant that can be used as a food source for humans, animals, and other living organisms. Additionally, corn can be used as an alternate energy source for vehicles and other technologies. However, using corn as an energy source has caused an increase in the cost of corn-based foods.

2. **Why was John Harvey Kellogg experimenting with cereal grains and doughs?**

   He was designing recipes that were easily digestible, vegetarian, and easy to prepare to feed the patients at the sanitarium where he worked.

3. **How were corn flakes invented?**

   Kellogg and his brother left boiled wheat out too long. When they tried to feed it through the rollers to make a thin layer of dough, the dough fell apart into small flakes. They baked these flakes and fed them to the patients.

Questions for Reflection

1. **Which recipe do you think would sell the best in a wide market? Why?**

   Student answers will vary.

2. **Which recipe do you think would be most appealing to children? Why? Which recipe would be most appealing to adults? Why?**

   Student answers will vary.

Apply and Analyze

1. **Why should food companies have food tasters? What is the importance of tasting foods before they go to market?**

   Food companies have food testers to examine the quality, taste, and appearance of food before it is introduced to the market. The food tests result in feedback that is used to improve the food in taste and/or appearance so that it will be more appealing to consumers.

2. **When Kirsten Hoskissen reviews cereals, what specifically does she observe and test to determine if the cereal is good for future sales?**

   The cereal is reviewed for taste, texture, color, smell, and appearance.

3. **What are potential challenges of being a food tester?**

   Students can name a variety of challenges, including food allergies and diet concerns (weight loss/gain, veganism). Testers also have to ignore their personal preferences in order to provide objective feedback.

Reflect

1. **What technologies might need to be developed to create or manufacture this design?**

   Answers may vary. However, if a student were to design a product that used corn to power vehicles, electronics, etc., he or she might cite the need for materials or technology that can transform the corn into energy.

2. **What are any constraints or drawbacks you can foresee with implementing this design?**

   Answers may vary. However, students might explain that the design or product developed would have to account for the sugar or energy stored in corn-based products and be adaptable to diverse or extreme environments.

3. **Would there be any environmental or human health concerns about using the product in this way?**

   Answers may vary. However, students might point out that corn-based products would need extensive testing to ensure safety for human/animal/environmental consumption.

Assessment

The Design Challenge can be assessed using the rubric in the appendix (p. 377).

Extensions

This lesson can be followed with lessons on chemistry-related properties of matter (physical and chemical properties) and energy transformation.

Resources and References

The Editors of Publications International, LDT. 9 things invented or discovered by accident. HowStuffWorks. *https://science.howstuffworks.com/innovation/scientific-experiments/9-things-invented-or-discovered-by-accident1.htm.*

Insley, J. 2010. A working life: The food taster. TheGuardian.com. *www.theguardian.com/money/2010/jun/19/working-life-food-taster.*

Markel, H. 2017. The secret ingredient in Kellogg's Corn Flakes is Seventh-Day Adventism. Smithsonian.com. *www.smithsonianmag.com/history/secret-ingredient-kelloggs-corn-flakes-seventh-day-adventism-180964247.*

Neal, A. 2015. 18 surprising, everyday items made with corn. Cheat Sheet. *www.cheatsheet.com/life/18-surprising-everyday-items-made-with-corn.html/?a=viewall.*

Parkin, B. *The Wall Street Journal.* 2017. Sneakers made from corn? Seat cushions from soybeans? May 14. *www.wsj.com/articles/sneakers-made-from-corn-seat-cushions-from-soybeans-1494813781.*

Smith, R. 2015. America's immortal cereal: The weird, wonderful story behind Corn Flakes, an *Object Lesson.* TheAtlantic.com. *www.theatlantic.com/health/archive/2015/04/americas-immortal-cereal/388991.*

GET FIRED UP WITH FRICTION LIGHTS

Matches

8

A Case Study Using the Discovery Engineering Process

Introduction

The ancestors of modern humans started using fire a million years ago. But prior to the 1820s, starting a fire could be a tedious and time-consuming process. Matches, or friction lights as they were once called, were the first reliable way to start a fire by rubbing two objects together. Today, matches are typically made of wooden or cardboard sticks and have one end coated with an ignitable material.

Lesson Objectives

By the end of this case study, you will be able to

- Describe the differences between friction lights and safety matches.

- Explain the differences between physical and chemical changes.

- Design an application in which a new kind of match solves a problem.

The Case

Read the following summary of the history and science of matches, and then answer the questions that follow.

In ancient times, humans started fire by rubbing together pieces of wood, stone, or metal to create friction and pressure. One day in 1825, John Walker, a chemist and pharmacist from England, was at home mixing together chemicals (paste of sulfur with gum, potassium chlorate, sugar, antimony trisulfide) with a stick. Then he scraped the chemical-coated stick across his hearth in order to clean it. As he did so, the chemicals burst into flame. Walker used the discovery to create what's known as friction lights or friction matches. Each match featured a stick with a ball of the aforementioned chemicals at one end.

Although Walker originally used cardboard to create his friction lights, he later switched to wooden splints and packaged the matches in a cardboard box with a piece of sandpaper to strike. Walker was able to sell his friction lights, but people felt that they were dangerous. The ball of chemicals often separated from the match and the burning head of the stick sometimes fell to the floor, causing damage. They would sometimes even self-ignite!

Walker did not patent his invention, and in 1829 Samuel Jones began making an exact copy of Walker's product, which he named *lucifer matches*. In the 1840s, a scientific breakthrough made by Gustaf Erik Pasch and Johan Edvard Lundstrom led to safety matches, where one chemical is included on the side of the matchbox rather than in the match head. Pasch and Lundstrom used red phosphorus on a specially designed igniting surface to ignite the match, decreasing the ability of matches to self-ignite.

Today, various types of matches can be found, including safety matches, strike-anywhere matches, and storm matches (Figure 8.1). Strike-anywhere matches are designed to ignite when they are drawn across any rough, dry surface. Storm matches are designed to be windproof and waterproof and to burn for a longer period of time. These matches can even reignite after quickly being immersed underwater. Storm matches are often included in survival kits.

FIGURE 8.1

Storm Matches

Recognize, Recall, and Reflect

1. What was John Walker doing when he discovered friction lights?

2. Why did people feel that friction lights were too dangerous?

3. What development made the matches safer?

Investigate

In this activity, you will explore chemical changes. Chemical changes (also called chemical reactions) are changes that produce new combinations of matter. Many chemical changes are irreversible. The burning of wood (like a lit match) results in a chemical change. A burning candle also demonstrates a chemical change. How so? Fire is a combustion reaction. Combustion reactions are chemical reactions that use oxygen and a carbon-based fuel (e.g., wood or evaporated wax) to create heat, light, and the byproducts of water and carbon dioxide. When you light a candle, it needs oxygen to continue burning. If you remove the oxygen, the fire will go out. To see this in action, watch a demonstration given by your teacher or form small groups and follow the directions below.

Materials

For each group of students or for the teacher demonstration:

- 3 tea candles

- 3 jars of various sizes

- Matches

- Stopwatch

- Safety glasses or goggles (1 pair per student)

Safety Note: Wear safety glasses or goggles during the setup, hands-on, and take-down segments of the activity. Use caution when working with matches and candles. These heat sources can seriously burn skin and clothing. Wash your hands with soap and water immediately after completing this activity.

Create, Innovate, and Investigate

- Label your different-size jars as 1, 2, or 3.

- Make a prediction for how long you think the candle will burn in each jar. Record your prediction in the data table below.

Jar	Estimated Time Candle Will Stay Lit	Actual Time Candle Stayed Lit	Other Observations
1			
2			
3			

- Light one candle and place the first jar over the candle. Don't forget to start the timer!

- Repeat the previous step for the next two candles.

Consider what you observed in the experiment and what you know about how lit candles work, then answer the questions below.

Questions for Reflection

1. What did you observe about the length of time each candle remained lit?

2. Did your prediction match your actual observations? Explain.

3. What can you conclude about how jar size affects the time a candle remains lit?

Apply and Analyze

In addition to demonstrating a chemical change, a lit candle can exhibit a physical change as well. Physical changes are changes in physical properties, such as when a solid melts into a liquid or when a liquid transitions into a gas. Physical changes are reversible. (For example, water can freeze into a solid [ice] and then melt back into liquid form.) A candle burning can demonstrate a physical change as the wax or paraffin melts and changes to a liquid, and the candle gets smaller.

Visit the following website to learn about gas, liquid, and solid states of matter (also known as phases of matter): *www.chem.purdue.edu/gchelp/atoms/states.html*. Then use what you have learned to answer the questions below.

1. What do you think happens to the particles in wax when a candle is lit? (Hint: Think about the different states of matter you read about.)

2. What are some other materials that change phase when you add or remove heat?

Design Challenge

Engineering is the application of scientific understanding through creativity, imagination, problem solving, and the designing and building of new materials to address and solve problems in the real world. You will be asked to take the science you have learned in this case and design a process or product to address a real-world issue of your choosing.

Engineers use the engineering design process as steps to address a real-world problem (see Figure 8.2). You will now use this process as you come up with a new type of matches. In this case, you are asking the question (Step 1) of how you can design a new type of matches. Drawing on your creativity, you will then brainstorm (Step 2) a new type of match that solves a problem. Afterward, you will create a plan (Step 3) for this new product. Next, you will create a sketch and/or model of your product (Step 4). Then, you will work with your classmates to think about how you would test (Step 5) and refine (Step 6) your product.

1. Ask Questions

Based on your prior knowledge, ask yourself what problems may be addressed or what products could be created by developing a new kind of match.

FIGURE 8.2

The Engineering Design Process

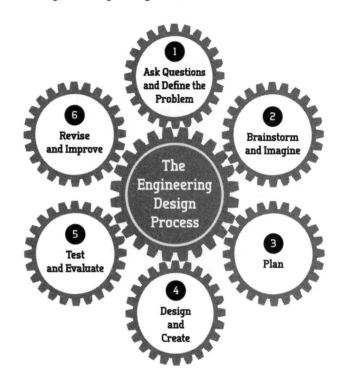

2. Brainstorm and Imagine

Brainstorm an application for a new kind of match. What problem will your product solve? (For example, the creation of matches that self-extinguish after five seconds could help people avoid burning their fingers on a lit match as the flame engulfs the matchstick.)

3. Create a Plan

Create a plan for your new product. Consider: (1) What is the purpose of the new product? (2) What are benefits of the product? (3) What are the limitations of the product? Use the Product Planning Graphic Organizer (p. 131) to help you.

4. Design and Create

Consider the following questions and considerations for your product and its design.

- How would your product benefit consumers?

- Are there any limitations or drawbacks to your match product? If so, how would you overcome them?

- What technologies might need to be developed to create or manufacture this design?

- What are any constraints or drawbacks you can foresee with implementing this design?

- Would there be any safety concerns regarding your product? If so, how would you address them?

Now, create a sketch of your product. Make sure your design incorporates the research and exploration you've done.

5. Test and Evaluate

Working with your classmates, come up with a way to test your design to see its effectiveness.

6. Revise and Improve

Give your plans to one of your classmates to review. Listen to his or her feedback on your design. What are some ways you can use the input to refine your design? Take some time to revise and make improvements.

Reflect

1. What technologies might need to be developed to create or manufacture this design?

2. What are any constraints or drawbacks you can foresee with implementing this design?

3. Would there be any environmental or human health concerns to using the product in this way?

Product Planning Graphic Organizer

Proposed Product Idea	
Pros (Benefits)	**Cons (Limitations)**

GET FIRED UP WITH FRICTION LIGHTS

MATCHES

A Case Study Using the Discovery Engineering Process

Lesson Overview

In this lesson, students explore matches and chemical reactions. Matches, also called friction lights, are a product inspired by an accidental discovery that made it much easier to start a fire.

Lesson Objectives

By the end of this case study, students will be able to

- Describe the differences between friction lights and safety matches.

- Explain the differences between physical and chemical changes.

- Design an application in which a new kind of match solves a problem.

The Case Study Approach

This lesson uses a case study approach. Explaining the purpose of case studies will encourage your students to relate to the material and engage with the problem. At the heart of each case study in this book is a true story, one that describes how someone in his or her everyday life or during a routine workday made an observation or did a simple experiment that led to a new insight or discovery. Case studies are designed to get students actively engaged in the process of problem solving. The narrative of the case supplies authentic details that place the student in the role of the inventor and provide scaffolds for critical thinking and deep reflection. A case is more than a paragraph to read or a story to analyze but rather a way of framing problems, synthesizing what is known, and thinking creatively about new applications and solutions. In this lesson, students consider how matches were discovered and work together to think about new applications for matches to solve real-life problems.

Use of the Case

Due to the nature of these case studies, teachers may elect to use any section of each case for their instructional needs. The sections are sequenced in order (scaffolded) so students think more deeply about the science involved in the case and develop an understanding of engineering in the context of science.

Curriculum Connections

Lesson Integration

You could use this case as a way to integrate engineering into lessons on chemical reactions, flammability, and phases of matter.

Related Next Generation Science Standards

PERFORMANCE EXPECTATIONS

- MS-PS1-2. Analyze and interpret data on the properties of substances before and after the substances interact to determine if a chemical reaction has occurred.

- HS-PS1-2. Construct and revise an explanation for the outcome of a simple chemical reaction based on the outermost electron states of atoms, trends in the periodic table, and knowledge of the patterns of chemical properties.

SCIENCE AND ENGINEERING PRACTICES

- Analyzing and Interpreting Data
- Constructing Explanations and Designing Solutions

CROSSCUTTING CONCEPTS

- Patterns
- Cause and Effect
- Energy and Matter

Related National Academy of Engineering Grand Challenge

- Engineer the Tools of Scientific Discovery

Lesson Preparation

You will need to make copies of the entire student section for the class. Students will need internet access at various points in the lesson. Alternatively, you can project videos or print and distribute copies of online content for the class. In the Investigate section, you can decide to have small groups of students carry out the experiment, or you can do the experiment as a class demonstration. If you decide to do a demonstration, perform the steps in the Create, Innovate, and Investigate section in front of the class. You can have students fill out the chart provided in this section as they watch the demonstration. Look at the Teaching Organizer (Table 8.1) for suggestions on how to organize the lesson.

Materials

For each group of students or for the teacher demonstration:

- 3 tea candles
- 3 jars of various sizes
- Matches
- Stopwatch
- Safety glasses or goggles (1 pair per student)

Safety Note for Students: Wear safety glasses or goggles during the setup, hands-on, and takedown segments of the activity. Use caution when working with matches and candles. These heat sources can seriously burn skin and clothing. Wash your hands with soap and water immediately after completing this activity.

Time Needed

55 minutes

Teaching Organizer

Section	Time Suggested	Materials Needed	Additional Considerations
The Case	5 minutes	Student packet	Could be read in class or as a homework assignment prior to class
Investigate	10 minutes	Student packet, 3 tea candles, 3 jars of various sizes, matches, stopwatch, safety glasses	Recommended as an in-class teacher demonstration or with students in groups testing their predictions
Apply and Analyze	10 minutes	Student packet, internet access	Small-group or individual activity
Design Challenge	30 minutes	Student packet, internet access	Small-group activity

Teacher Background Information

Students often have questions about physical and chemical changes.

Vocabulary

- chemical change
- flammable
- phases of matter

Teacher Answer Key

Recognize, Recall, and Reflect

1. **What was John Walker doing when he discovered friction lights?**

   Cleaning chemicals off a stick at his home by scraping the stick across his hearth.

2. **Why did people feel that friction lights were too dangerous?**

   The ball of chemicals at the top often separated from the matchstick and fell, causing damage. The matches could also occasionally self-ignite.

3. **What development made the matches safer?**

   The invention of safety matches, where one of the key chemicals is included on the side of the matchbox instead of in the match head, made matches safer as it decreased the chances of the matches self-igniting.

Questions for Reflection

1. **What did you observe about the length of time each candle remained lit?**

   Answers may vary. However, students should observe that the candles remained lit for various lengths of time depending on the jar size.

2. **Did your prediction match your actual observations? Explain.**

   Student answers will vary based on their initial predictions.

3. **What can you conclude about how jar size affects the time a candle remains lit?**

   The larger the jar, the more oxygen it contains. And because candles need oxygen to stay lit, a candle will burn longer under a larger jar than it will under a smaller one with less oxygen.

Apply and Analyze

1. **What do you think happens to the particles in wax when a candle is lit? (Hint: Think about the different states of matter you read about.)**

   The wax will melt, transforming from a solid into a liquid; that means the particles inside will go from being rigid to being able to slide past one another.

2. **What are some other materials that change phase when you add or remove heat?**

   Answers will vary but could include items that go through a physical change (e.g., water becoming ice cubes, water vapor becoming dew drops).

Reflect

1. **What technologies might need to be developed to create or manufacture this design?**

   Answers will vary based on the students' designs.

2. **What are any constraints or drawbacks you can foresee with implementing this design?**

   Answers will vary based on the students' designs. Examples could include safety as being a constraint for using chemical reactions.

3. **Would there be any environmental or human health concerns to using the product in this way?**

   Answers will vary based on the students' designs.

Assessment

Check for understanding of physical and chemical changes with a quiz, or ask students to write a formal lab report based on their findings from this experiment. The Design Challenge can be assessed using the rubric in the appendix (p. 377).

Extensions

This lesson can be followed with lessons about other types of reactions, chemical properties of elements, or the periodic table.

Resources and References

Brar, A. Batteries: Electricity through chemical reactions. LibreTexts. *https://chem.libretexts. org/Core/Analytical_Chemistry/Electrochemistry/Case_Studies/Batteries%3A_Electricity_ though_chemical_reactions* (accessed September 15, 2017).

The British Museum. A history of the world: John Walker's friction lights. BBC. *www.bbc. co.uk/ahistoryoftheworld/objects/hQR9oN5LTeCLcuKfPDMJ9A.*

History of Matches. Early and modern matches. *www.historyofmatches.com/matches-history/ history-of-matches* (accessed December 28, 2017).

McIntosh, J., S. White, and R. Suter. 2009. Science sampler: Enhancing student understanding of physical and chemical changes. *Science Scope* 33 (2): 54.

Miller, K. 2013. Archaeologists find earliest evidence of humans cooking with fire. *Discover Magazine. http://discovermagazine.com/2013/may/09-archaeologists-find-earliest-evidence-of- humans-cooking-with-fire*

Purdue University. States of matter. *www.chem.purdue.edu/gchelp/atoms/states.html* (accessed December 28, 2017).

THAT'S A WRAP

Plastic

A Case Study Using the Discovery Engineering Process

Introduction

Billions of people encounter plastics every day, often in the form of water bottles, toys, food packaging, and utensils. Did you know that synthetic plastic was first created in 1905? Let's explore how plastic became such a common product.

Lesson Objectives

By the end of this case study, you will be able to

- Describe how plastic was discovered.

- Analyze the benefits and limitations of Bakelite.

- Design a new application for plastic polymers that solves a problem.

The Case

Read the following summary of the discovery and history of plastics.

In 1905, Belgian American chemist Leo Baekeland was interested in creating a synthetic substitute for shellac. Shellac is a commercial resin made from secretions of the female lac bug found in India and Thailand. It is a natural thermoplastic, which becomes soft and flowing when heated. Shellac has been used in various ways, including as an electrical insulator. However, it takes some 15,000 beetles about six months to produce one pound of the resin.

While conducting research to find a shellac substitute, Baekeland discovered the first entirely synthetic plastic and called it Bakelite. Formed at high temperature and pressure, Bakelite is a product of a condensation reaction of formaldehyde and phenol. The material is known as a thermoset plastic, which means that its polymer chains are connected through heat and pressure in the molding process; and once they are linked, they cannot be unlinked. Bakelite can be broken into bits but not easily melted or recycled. (Polymers are molecules consisting of repeating subunits called monomers.)

Bakelite has many of the same properties as other plastics made today. It is lightweight, durable, and easily molded—a benefit in the mass production of products. Bakelite also has low conductivity (the ability to transmit electricity), which made it very valuable in the electrical and automotive industries at the time of its discovery. When the Ericsson company began using Bakelite to make its telephones in the 1930s (Figure 9.1), the time needed to make a phone casting was reportedly reduced from one week to about seven minutes.

Bakelite jewelry became popular during the Art Deco period (1909 to the 1940s). It was used as costume jewelry and was a cheap way to accessorize. Bakelite production was stalled during World War II due to a need for war-related product production. By the end of the war, more advances in plastics had been developed, decreasing the interest in Bakelite.

FIGURE 9.1

Bakelite Telephone

Bakelite was once used for nonconducting parts of phones.

Recognize, Recall, and Reflect

1. Why was Baekeland searching for a synthetic shellac substitute?

2. What is thermoset plastic?

3. What happened to stall the production of Bakelite?

Investigate

In this activity, you will explore plastics that are not thermoset. Plastic #6 (the clear plastic used in food containers) is an example of a thermoplastic that has been heated, stretched, and cooled into shape. If heated again, this plastic will go back to its original state.

Materials

For each student:

- Permanent markers in a variety of colors

- Access to an oven or toaster oven

- Cookie sheet or baking tray (students can share sheets or trays)

- Parchment paper or aluminum foil to cover tray

- Potholder

- #6 plastic (e.g., clear takeout food containers or polymer craft sheets)

- Scissors

- Hole punch

- Safety glasses or goggles

- Heat-resistant gloves

Safety Note: Precautions must be taken when melting your plastic using the oven. Listen to instructions from your teacher. Wear safety glasses or goggles during the setup, hands-on, and takedown segments of the activity. Use caution when working with hot toaster ovens or other heat sources; these can cause skin burns or electric shock. Have good ventilation when heating plastic to reduce exposure to vapors. Use caution when using sharp tools; these can cut or puncture skin. Wash your hands with soap and water immediately after completing this activity.

Create, Innovate, and Investigate

- Begin by observing your piece of plastic. Does it have any unusual properties?

- Next, decorate your piece of plastic with the permanent markers. You can use the scissors to cut your plastic into a shape. If you want to hang your creation when you are done, use the hole punch to create a hole in the material before heating your plastic.

- Once the oven has heated to 350°F, work with your teacher to put your plastic piece onto a cookie sheet or tray and place it inside the oven. The cookie sheet or tray should be lined with parchment paper or aluminum foil so your plastic does not stick.

- Wait two to three minutes for your plastic to shrink. (Once the process begins, it happens quickly!)

- Working with your teacher and using a potholder, remove the cookie sheet or tray and wait five minutes for the plastic to cool completely before touching it.

Questions for Reflection

1. What did you observe about your creation?

2. When you heat the plastic, how does it behave differently?

3. What can you conclude about #6 plastics?

Apply and Analyze

Plastic technology is quickly being adopted for new areas of research and new commercial applications. For instance, researchers at North Carolina State University have developed a way to fold polymer sheets (the same type of plastic you used during your investigation) using only light. Check out the following video explaining the process of self-folding polymer sheets: *www.youtube.com/watch?v=ZlZOdiwbZIE*. Then watch a video that shows how light can turn 2-D plastic patterns into 3-D shapes: *www.youtube.com/watch?v=O9WnvixaZRQ*. After exploring each video, answer the questions that follow.

1. What is a benefit of folding a polymer sheet using only light?

2. What are some new applications that could be explored using this technology?

Design Challenge

Engineering is the application of scientific understanding through creativity, imagination, problem solving, and the designing and building of new materials to address and solve problems in the real world. You will be asked to take the science you have learned in this case and design a process or product to address a real-world issue of your choosing.

Engineers use the engineering design process as steps to address a real-world problem (see Figure 9.2). You will now use this process as you come up with a new application for plastic. In this case, you are asking the question (Step 1) of how plastic could be used for new purposes. Drawing on your creativity, you will then imagine (Step 2) a new plastic product that will solve a problem. Next, you will create a plan (Step 3) for this new product. Although you will not actually produce your product, you will create a sketch and/or a model of it (Step 4). Then, you will work with your classmates to think about how you would test (Step 5) and refine (Step 6) your product.

FIGURE 9.2

The Engineering Design Process

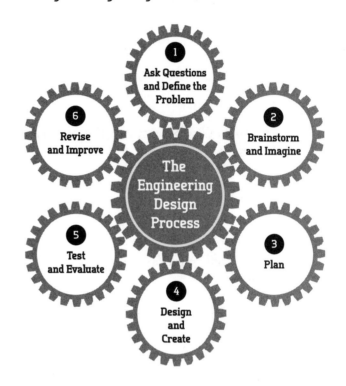

1. Ask Questions

Consider a new problem that may be addressed or a product that could be created by using plastic polymers. What are some applications in which you would need a material that becomes flexible when heated? What problem could be solved with this?

2. Brainstorm and Imagine

Based on your previous research, imagine a new application for plastic polymers. What useful products could be made using this technology? (For example, one idea is to use plastic polymers to self-seal packaging.)

3. Create a Plan

Create a plan for a new plastic-polymer product. Consider: (1) What is the purpose of the product? (2) What are benefits of the product? (3) What are the limitations of the product? Use the Product Planning Graphic Organizer (p. 145) to help you.

4. Design and Create

Consider the following questions and considerations for your plastic polymer product and its design.

- How would incorporating plastic polymers into your product design make it better?

- Are there any limitations or drawbacks to using plastic polymers in your product? If so, how would you overcome them?

- What technologies might need to be developed to create or manufacture this design?

- What are any constraints or drawbacks you can foresee with implementing this design?

- Would there be any safety concerns regarding your plastic polymer–based product?

Now, create a sketch of your plastic-polymer product. Make sure your design incorporates the research and exploration you've done.

5. Test and Evaluate

Working with your classmates, come up with a way to test your design to see its effectiveness.

6. Revise and Improve

Give your plans to one of your classmates for review. Listen to his or her feedback on your design. What are some ways you can use the input to refine your design? Take some time to revise and make improvements.

Reflect

1. What technologies might need to be developed to create or manufacture this design?

2. What are any constraints or drawbacks you can foresee with implementing this design?

3. Would there be any environmental or human health concerns to using the product in this way?

Product Planning Graphic Organizer

Proposed Product Idea	
Pros (Benefits)	**Cons (Limitations)**

THAT'S A WRAP

PLASTIC

A Case Study Using the Discovery Engineering Process

Lesson Overview

In this lesson, students explore plastic polymers. Plastic was an accidental discovery that has led to a number of new applications and products.

Lesson Objectives

By the end of this lesson, students will be able to

- Describe how plastic was discovered.

- Analyze the benefits and limitations of Bakelite.

- Design a new application for plastic polymers that solves a problem.

The Case Study Approach

This lesson uses a case study approach. Explaining the purpose of case studies will encourage your students to relate to the material and engage with the problem. At the heart of each case study in this book is a true story, one that describes how someone in his or her everyday life or during a routine workday made an observation or did a simple experiment that led to a new insight or discovery. Case studies are designed to get students actively engaged in the process of problem solving. The narrative of the case supplies authentic details that place the student in the role of the inventor and provide scaffolds for critical thinking and deep reflection. A case is more than a paragraph to read or a story to analyze but rather a way of framing problems, synthesizing what is known, and thinking creatively about new applications and solutions. In this lesson, students consider how plastic was discovered and work together to think about new applications for plastic to solve real-life problems.

Use of the Case

Due to the nature of these case studies, teachers may elect to use any section of each case for their instructional needs. The sections are sequenced in order (scaffolded) so students think more deeply about the science involved in the case and develop an understanding of engineering in the context of science.

Curriculum Connections

Lesson Integration

You could use this case as a way to integrate engineering into a lesson on atomic structure, chemistry, or metals and the periodic table.

Related Next Generation Science Standards

PERFORMANCE EXPECTATIONS

- HS-PS2-6. Communicate scientific and technical information about why the molecular-level structure is important in the functioning of designed materials.

- HS-ETS1-2. Design a solution to a complex real-world problem by breaking it down into smaller, more manageable problems that can be solved through engineering.

- HS-ETS1-3. Evaluate a solution to a complex real-world problem based on prioritized criteria and trade-offs that account for a range of constraints, including cost, safety, reliability, and aesthetics, as well as possible social, cultural, and environmental impacts.

SCIENCE AND ENGINEERING PRACTICES

- Analyzing and Interpreting Data
- Constructing Explanations and Designing Solutions

CROSSCUTTING CONCEPT

- Structure and Function

Related National Academy of Engineering Grand Challenge

- Engineer the Tools of Scientific Discovery

Lesson Preparation

Look at the Teaching Organizer (Table 9.1) for suggestions on how to organize the lesson. Note that you will need to make copies of the entire student section for the class.

During the Investigation section, students can use scissors to cut their plastic container or craft sheet. Punching a hole ahead of baking the creations will allow students to hang up their final products. You will need access to an oven or toaster oven, which you must preheat to 350°F. The process usually takes two to five minutes, depending on the oven. You may want to use disposable aluminum trays for this activity. (We strongly discourage reusing these trays for cooking food as you might risk mixing leftover plastic with food materials.) Be sure to cover each cookie sheet or baking tray with parchment paper or aluminum foil so the plastic does not stick. If the oven is in a confined space, make sure the area is properly ventilated. Also ensure students' safety when handling hot cookie sheets or baking trays. Allow plastic to cool completely before touching.

Access to the internet is required for the Apply and Analyze portion of the lesson as there are videos for students to view. If access is unavailable for individual students, you could set up a projector to display the videos.

Materials

For each student:

- Permanent markers in a variety of colors

- Access to an oven or toaster oven

- Cookie sheets or baking trays (students can share sheets or trays)

- Parchment paper or aluminum foil to cover trays

- Potholder

- #6 plastic (e.g., clear takeout food containers or polymer craft sheets)

- Scissors

- Hole punch

- Safety glasses or goggles

- Heat resistant-gloves

Safety Note for Students: Precautions must be taken when melting your plastic using the oven. Listen to instructions from your teacher. Wear safety glasses or goggles during the setup, hands-on, and takedown segments of the activity. Use caution when working with hot toaster ovens or other heat sources; these can cause skin burns or electric shock. Have good ventilation when heating plastic to reduce exposure to vapors. Use caution when using sharp tools; these can cut or puncture skin. Wash your hands with soap and water immediately after completing this activity.

Time Needed

55 minutes

TABLE 9.1

Teaching Organizer

Section	Time Suggested	Materials Needed	Additional Considerations
The Case	5 minutes	Student packet	Could be read in class or as a homework assignment prior to class
Investigate	10 minutes	Student packet, permanent markers in a variety of colors, access to an oven or toaster oven, cookie sheets or baking trays, parchment paper or aluminum foil to cover trays, potholder, #6 plastic (e.g., clear takeout food containers or polymer craft sheets), scissors, hole punch, safety glasses or goggles, heat-resistant gloves	Students will work individually to make their plastic creations, but all students will need supervision when using the oven.
Apply and Analyze	10 minutes	Student packet, internet access	Small-group or individual activity
Design Challenge	30 minutes	Student packet	Small-group activity

Teacher Background Information

There are a number of resources and videos about plastics and polymers available on the internet. You may want to observe the behavior of plastics on sites such as YouTube prior to using the case.

Vocabulary

- formaldehyde
- phenol
- polymer
- resin
- synthetic
- thermoset

Teacher Answer Key

Recognize, Recall, and Reflect

1. **Why was Baekeland searching for a synthetic shellac substitute?**

   To avoid using beetles to produce the resin

2. **What is thermoset plastic?**

   It's a material with polymer chains that are hooked together through the heat and pressure applied when it is molded. Once linked, these chains cannot be unlinked.

3. **What happened to stall the production of Bakelite?**

   World War II and developments in different plastics

Questions for Reflection

1. **What did you observe about your creation?**

   Student creations will shrink and the plastic will become flat. The color intensifies after it shrinks.

2. **When you heat the plastic, how does it behave differently?**

   The plastic will first curl up, but then it will flatten back out.

3. **What can you conclude about #6 plastics?**

   They shrink when exposed to high heat and become hard.

Apply and Analyze

1. **What is a benefit of folding a polymer sheet using only light?**

   Answers may vary but could include not having to touch the material, etc.

2. **What are some new applications being explored using this technology?**

   Packaging

Reflect

1. **What technologies might need to be developed to create or manufacture this design?**

   Answers will vary based on student designs.

2. **What are any constraints or drawbacks you can foresee with implementing this design?**

   Answers will vary based on student designs.

3. **Would there be any environmental or human health concerns to using the product in this way?**

   Answers will vary based on student designs but could include issues with plastic waste disposal.

Assessment

The Design Challenge can be assessed using the rubric in the appendix (p. 377).

Extensions

This lesson can be followed by lessons about chemical reactions, chemical properties, or synthetic materials. Encourage students to investigate self-assembly in biological systems (like DNA transcription) and engineered products. Students could also research plastic waste and how to solve the problem of plastics in our oceans.

Resources and References

American Chemistry Council. The basics: Polymer definition and properties. *https://plastics. americanchemistry.com/plastics/The-Basics* (accessed January 3, 2018).

Dickey, M. 2012. "Self-folding origami of plastic sheets." YouTube video. *www.youtube. com/watch?v=O9WnvixaZRQ.*

Dickey, M. 2017. "Sequential self-folding of polymer sheets." YouTube video. *www. youtube.com/watch?v=ZlZOdiwbZIE.*

Ericsson. The Bakelite telephone 1931. *www.ericsson.com/en/about-us/history/products/the-telephones/the-bakelite-telephone-1931* (accessed January 9, 2018).

Freinkel, S. 2011. A brief history of plastic's conquest of the world. *Scientific American. www. scientificamerican.com/article/a-brief-history-of-plastic-world-conquest* (accessed January 9, 2018).

Liu, Y., J. K. Boyles, J. Genzer, and M. D. Dickey. 2012. Self-folding of polymer sheets using local light absorption. *Soft Matter* 8 (6): 1764–1769.

Plastics Make It Possible. 2012. Bakelite: The plastic that made history. *www. plasticsmakeitpossible.com/whats-new-cool/fashion/styles-trends/bakelite-the-plastic-that-made-history* (accessed January 9, 2018).

Powers, V. 1993. *The Bakelizer.* American Chemical Society. *www.acs.org/content/acs/en/ education/whatischemistry/landmarks/bakelite.html.*

Teeda. History of Bakelite jewelry. *www.teeda.com/pages/history-of-bakelite-jewelry* (accessed January 9, 2018).

A STICKY DISCOVERY

The Invention of Post-It Notes

> ## A Case Study Using the Discovery Engineering Process

Introduction

In this lesson, you will learn about the invention of Post-it notes. You will explore what makes these handy little notes sticky and invent a new product that uses a similar type of adhesive.

Lesson Objectives

By the end of this case study, you will be able to

- Describe how sticky notes work.

- Analyze how long sticky notes will effectively adhere.

- Design a new product that incorporates a gentle adhesive.

The Case

Read the following summary of the invention of Post-it notes.

Post-it notes owe their existence to a discovery made back in 1968 by Dr. Spencer Silver, a 3M Company researcher who was trying to create newer and stronger adhesives. But instead of developing a stronger glue, Silver discovered a glue that was weak and allowed two glued surfaces to separate with ease. For years, Silver

could not find a good use for his discovery. But he kept talking about it with his colleagues.

As legend has it, another 3M scientist named Arthur Fry noticed that all his bookmarks in his choir hymnals kept falling out. Fry wanted to find a bookmark that would stay in place but not damage the hymnal. He remembered a talk that Silver had given and realized that the gentle glue might be the perfect solution. Fry worked with Silver to develop what we now know as the Post-it note.

The two researchers used the notes around the office and people realized just how useful this new product was. Not only were the notes gentle when they stuck to surfaces, but they could also be reused. The company tried the Post-it notes out in a test market in Boise, Idaho. Everyone who used the product said they would buy it if it were to become available. After this trial, the product became a huge success.

Recognize, Recall, and Reflect

1. What was Spencer Silver trying to invent when the glue for Post-it notes was discovered?

2. What everyday problem prompted the invention of Post-it notes?

3. What was unique about Post-it notes compared to other notepads or types of adhesives?

Investigate

How strong is a sticky note? One of the reasons these products are so popular is that they are strong enough to stick to a surface without falling off. In this activity, you will examine the adhesiveness of these types of notes and explore different sizes of sticky notes to see if size makes a difference in how well they stick. First, you will need to create a system to test the strength of each note.

(Hint: One easy way to test the strength of each sticky note is to attach a binder clip to the bottom of the note. Then stick the note on the vertical edge of your desk or to a bookcase. Finally, place paperclips or small weights onto the wire handles of the binder clip to see how much weight your note will hold [Figure 10.1]. Design a test to determine if small notes are stronger than large notes. Does the amount of sticky surface matter?)

Materials

For each group of students:

- 3 different sizes of sticky notes

- Materials with different textures (desks, books, fabrics)

- Binder clips

- Paperclips or small weights

- Safety glasses or goggles (1 pair per student)

Create, Innovate, and Investigate

- Begin by observing how sticky notes stick to different surfaces. What do you observe about the way the paper adheres to each surface?

- Try sticking a sticky note on your desk multiple times. Does it lose its adhesive properties?

- Experiment with different sizes of sticky notes. Does the size of the note change how effective it is at sticking to a surface?

- What questions do you have about how sticky notes work?

Investigation Design

Questions for Reflection

1. What did you observe about the properties of sticky notes?

2. What are the unusual properties of sticky notes?

3. Do different sizes of sticky notes behave differently?

Apply and Analyze

Read the information and explore the links that follow to answer the questions below.

Throughout history, people have searched for effective adhesives that can bond two surfaces together. Early glues were made from materials such as fish skin or other animal products. How do adhesives work? What makes one strong and another weak?

There are many reasons why things stick together. Geckos can stick to smooth surfaces without any help from glue, hooks, or even little suction cups. Instead, a gecko's toe pads are covered in tiny, hairlike projections known as setae. These projections have an intermolecular (electrostatic) attraction to smooth surfaces like glass. Moreover, the setae have such a large surface area that the intermolecular attraction is strong enough to allow the gecko to hold onto the glass. This type of molecular attraction is known as van der Waals forces.

Some glues work by creating a chemical bond that forms a new chemical compound to bind materials together. An example of this type of adhesive is superglue, which forms a plastic mesh when it comes into contact with water molecules in the air.

Other adhesives work with a mechanical bonding where the adhesive enters tiny pores in materials, allowing two materials to stick together as if tiny screws were used. Products like Velcro work through a mechanical kind of adhesion, using hooks and loops.

Materials like packaging tape work through cohesion and adhesion. Cohesive forces bind two similar molecules together. Adhesive forces bind unlike molecules. For example, water molecules bond to each other through hydrogen bonding (cohesion) and a water droplet will stick to a sheet of glass (adhesion). Packaging tape works by having pressure-sensitive glue that has elastic properties and, when pressed onto a box, it stretches and will adhere to the box. The pressure-sensitive glue is viscous (it remains tacky) at a range of temperatures and, when pressure is applied, the glue wets the surface and adheres.

Sticky notes work in ways that are similar to packaging tape. These notes feature a thin layer of pressure-sensitive adhesive that includes tiny microcapsules of glue. If examined under a scanning electron microscope, the glue layer looks bumpy—similar to the surface of a basketball. When you stick a sticky note on a table, lift it up, and then stick it down again, some of these microcapsules pick up dirt particles and some are left behind on the table. This is why sticky notes can only be used a few times before they weaken and get covered with dust and dirt.

Since the invention of sticky notes, people have come up with lots of different ways to use the product. Sticky notes have emerged as a great way to label everything from electrical cables to file folders to storage boxes. These notes can also be used for to-do lists and art projects. Visit the websites below to learn more.

- *www.adhesives.org/adhesives-sealants/science-of-adhesion/why-does-an-adhesive-bond*

- *www.scientificamerican.com/article/what-exactly-is-the-physi*

1. How are geckos able to stick to glass surfaces?

2. How does superglue work differently from pressure-sensitive adhesives?

3. Does a sticky note last forever? What happens to it over time?

Design Challenge

Engineering is the application of scientific understanding through creativity, imagination, problem solving, and the designing and building of new materials to address and solve problems in the real world. You will be asked to take the science you have learned in this case and design a process or product to address a real-world issue of your choosing.

Engineers use the engineering design process as steps to address a real-world problem (see Figure 10.2). You will now use this process as you come up with a new application for a gentle adhesive. In this case, you are asking the question (Step 1) of how you can design a new use for gentle adhesives. Drawing on your creativity, you will then brainstorm

FIGURE 10.2

The Engineering Design Process

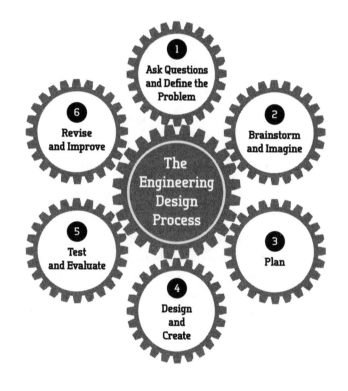

(Step 2) a specific product that uses a gentle adhesive to solve a problem. Next, you will create a plan (Step 3) for your product. Although you will not actually produce your product, you will create a sketch and/or a model of it (Step 4). Then, you will work with your classmates to think about how you would test (Step 5) and refine (Step 6) your product.

1. Ask Questions

Based on your research, consider a new problem that may be addressed or a product that could be created by using a new kind of adhesive. What are some applications where you would need a material that could be gently removed? What problem could be solved with such an adhesive?

2. Brainstorm and Imagine

Use what you read in the Apply and Analyze section to help you brainstorm a new application for a gentle adhesive. What useful products could be made using this technology? (For example, imagine an adhesive used for hanging holiday lights that would break down after a set number of days or months. With this product, you could put up the lights on, say, December 1, and they would gently fall down on January 1. Or perhaps you could design easy-to-remove adhesive bandages or hair clips that gently stick to hair.)

3. Create a Plan

Create a plan for your product. Consider: (1) What is the purpose of the product? (2) What are benefits to using this product? (3) What are the limitations of using this product? Use the Product Planning Graphic Organizer (p. 160) to help you.

4. Design and Create

Consider the following questions and considerations for your gentle adhesive product and its design.

- How would the product work?

- What technologies might need to be developed to create or manufacture this design?

- What are any constraints or drawbacks you can foresee with implementing this new product design? (A number of environmental waste products result from the production of sticky notes, including waste from making paper, waste from making glue, waste from packaging and distribution, and waste from people using Post-its. Some of the chemicals that may result from the production of sticky notes include caustic acids, carbon dioxide, mercury, Carbopol, and Rhodasurf. Think about how to avoid some of these waste products as you design your new product.)

- Would there be any safety concerns regarding your gentle adhesive product?

Now, create a sketch of your product. Make sure your design incorporates your research and exploration you've done.

5. Test and Evaluate

Working with your classmates, come up with a way to test your design to see its effectiveness.

6. Revise and Improve

Give your plans to one of your classmates for review. Listen to his or her feedback on your design. What are some ways you can use the input to refine your design? Take some time to revise and make improvements.

Reflect

1. What technologies might need to be developed to create or manufacture this design?

2. What are any constraints or drawbacks you can foresee with implementing this design?

3. Would there be any environmental or human health concerns to using the product in this way?

Product Planning Graphic Organizer

Proposed Product Idea	
Pros (Benefits)	**Cons (Limitations)**

A STICKY DISCOVERY

THE INVENTION OF POST-IT NOTES

A Case Study Using the Discovery Engineering Process

Lesson Overview

In this lesson, students explore properties and characteristics of Post-it notes and other sticky notes. Post-it notes resulted from an accidental discovery that has led to them being a household item. Students also learn about the properties of other adhesives and design their own product that incorporates a gentle adhesive to solve a problem.

Lesson Objectives

By the end of this case study, students will be able to

- Describe how sticky notes work.
- Analyze how long sticky notes will effectively adhere.
- Design a new product that incorporates a gentle adhesive.

The Case Study Approach

This lesson uses a case study approach. Explaining the purpose of case studies will encourage your students to relate to the material and engage with the problem. At the heart of each case study in this book is a true story, one that describes how someone in his or her everyday life or during a routine workday made an observation or did a simple experiment that led to a new insight or discovery. Case studies are designed to get students actively engaged in the process of problem solving. The narrative of the case supplies authentic details that place the student in the role of the inventor and provide scaffolds for critical thinking and deep reflection. A case is more than a paragraph to read or a story to analyze but rather a way of framing problems, synthesizing what is known, and thinking creatively about new applications and solutions. In this lesson, students consider how Post-it notes were discovered and work together to think about new applications for gentle adhesives to solve real-life problems.

Use of the Case

Due to the nature of these case studies, teachers may elect to use any section of each case for their instructional needs. The sections are sequenced in order (scaffolded) so students think more deeply about the science involved in the case and develop an understanding of engineering in the context of science.

Curriculum Connections

Lesson Integration

You could use this case as a way to integrate engineering into a lesson on atomic structure, bonds, and cohesion and adhesion. This case will work well with Chapter 5, "A Sticky Solution: Gecko Feet Adhesives" (p. 73).

Related Next Generation Science Standards

PERFORMANCE EXPECTATIONS

- HS-PS2-6. Communicate scientific and technical information about why the molecular-level structure is important in the functioning of designed materials.

- HS-ETS1-2. Design a solution to a complex real-world problem by breaking it down into smaller, more manageable problems that can be solved through engineering.

- HS-ETS1-3. Evaluate a solution to a complex real-world problem based on prioritized criteria and trade-offs that account for a range of constraints, including cost, safety, reliability, and aesthetics, as well as possible social, cultural, and environmental impacts.

SCIENCE AND ENGINEERING PRACTICES

- Analyzing and Interpreting Data
- Engaging in Argument From Evidence
- Constructing Explanations and Designing Solutions

Related National Academy of Engineering Grand Challenge

- Engineer the Tools of Scientific Discovery

Lesson Preparation

You will need to make copies of the entire student section for the class. Students will need internet access at various points in the lesson. Alternatively, you can project videos or print and distribute copies of online content for the class. For the Investigate section, you may want to create a model system for examining the strength of sticky notes to show to the class as an example. There are several interesting examples online of art created from sticky notes. Showing these examples to students is one way to introduce or end the lesson. Look at the Teaching Organizer (Table 10.1) for suggestions on how to organize the lesson.

Materials

For each group of students:

- 3 different sizes of sticky notes

- Materials with different textures (desks, books, fabrics)

- Binder clips

- Paperclips or small weights

- Safety glasses or goggles (1 pair per student)

Time Needed

65 minutes

TABLE 10.1

Teaching Organizer

Section	Time Suggested	Materials Needed	Additional Considerations
The Case	5 minutes	Student packet	Could be read in class or as a homework assignment prior to class
Investigate	15 minutes	Student packet, 3 different sizes of sticky notes, materials with different textures (desk, book, fabric), binder clips, paperclips or small weights, safety glasses or goggles	Recommended as a small-group activity
Apply and Analyze	15 minutes	Student packet, internet access	Whole-class or individual activity
Design Challenge	30 minutes	Student packet	Small-group activity

Teacher Background Information

Students may need a review of cohesion and adhesion. Khan Academy has a web page that might be helpful in teaching these concepts as they relate to water: *www.khanacademy.org/science/biology/water-acids-and-bases/cohesion-and-adhesion/a/cohesion-and-adhesion-in-water.*

Vocabulary

- adhesion
- adhesive
- cohesion
- compound

- elasticity
- electrostatic attraction
- molecule
- van der Waals forces

Teacher Answer Key

Recognize, Recall, and Reflect

1. **What was Spencer Silver trying to invent when the glue for Post-it notes was discovered?**

   He was trying to create a stronger adhesive.

2. **What everyday problem prompted the invention of Post-it notes?**

   There was a need for labels and placeholders that were not permanent and did not damage a surface when removed.

3. **What was unique about Post-it notes compared to other notepads or types of adhesives?**

   The unique thing about Post-it notes was that they could stick firmly to paper and be removed without destroying the paper. They could also be reused.

Questions for Reflection

1. **What did you observe about the properties of sticky notes?**

   Student answers will vary.

2. **What are the unusual properties of sticky notes?**

   The notes can easily be removed and used again.

3. **Do different sizes of sticky notes behave differently?**

No, different-size notes behave similarly.

Apply and Analyze

1. **How are geckos able to stick to glass surfaces?**

The hairlike setae on a gecko's toe pads have an intermolecular (electrostatic) attraction to smooth surfaces like glass. This molecular attraction is known as van der Waals forces.

2. **How does superglue work differently from pressure-sensitive adhesives?**

Superglue works through a chemical reaction. It creates a chemical bond that forms a new chemical compound to bind materials together. Pressure-sensitive adhesives don't create a chemical reaction. Instead, they have elastic properties and, when pressed onto a surface, they stretch and adhere to that surface.

3. **Does a sticky note last forever? What happens to it over time?**

Sticky notes will pick up dirt and debris. After a while they do not stick as well.

Reflect

1. **What technologies might need to be developed to create or manufacture this design?**

Students might discuss technology needed to cut, assemble, and package their designs.

2. **What are any constraints or drawbacks you can foresee with implementing this design?**

Students might discuss the waste that could result from the manufacturing process.

3. **Would there be any environmental or human health concerns to using the product in this way?**

Students' answers will vary, but they should consider whether chemicals used to make their products might be harmful.

Assessment

The Design Challenge can be assessed using the rubric in the appendix (p. 377).

Extensions

This lesson can be paired with the case study in Chapter 17, "Super Glue: Accidentally Discovered Twice" (p. 269). It could also be taught with Chapter 22, "Velcro: Engineering Mimics Nature" (p. 363). You can dive deeper into the nature of sticky note adhesives by having students investigate the number of times a sticky note will restick to a surface or the way in which notes stick to different wall textures.

Resources and References

Adhesives.org. Why does an adhesive bond? *www.adhesives.org/adhesives-sealants/science-of-adhesion/why-does-an-adhesive-bond* (accessed January 12, 2018).

EepyBird. The extreme sticky note experiments. *www.eepybird.com/featured-video/the-extreme-sticky-note-experiments* (accessed January 12, 2018).

Halford, B. 2004. Sticky notes: Serendipitous chemical discovery and a bright idea led to a new product that is ubiquitous. *Chemical & Engineering News* 82 (14): 64. *http://pubs.acs.org/cen/whatstuff/stuff/8214sci3.html.*

Khan Academy. Cohesion and adhesion of water. *www.khanacademy.org/science/biology/water-acids-and-bases/cohesion-and-adhesion/a/cohesion-and-adhesion-in-water* (accessed January 12, 2018).

ScientificAmerican.com. What exactly is the physical or chemical process that makes adhesive tape sticky? *www.scientificamerican.com/article/what-exactly-is-the-physi* (accessed January 12, 2018).

Tenner, E. 2010. The dark side of sticky notes. *The Atlantic. www.theatlantic.com/technology/archive/2010/07/the-dark-side-of-sticky-notes/60543.*

PUTTY IN YOUR HANDS

The Unintended Discovery of Silly Putty

A Case Study Using the Discovery
Engineering Process

Introduction

If you have ever played with Silly Putty,
you know that it is a doughlike toy
product with some unexpected properties. It bounces like a ball, but it can also
float in liquids or form a puddle on a
tabletop. Toy putty is a non-Newtonian
fluid, which means it acts like both a
liquid and a solid.

Lesson Objectives

By the end of this case study, you will
be able to

- Describe the properties of a non-Newtonian fluid.

- Analyze the behavior of toy putty.

- Design a new product or application that uses toy putty or another non-
  Newtonian fluid.

The Case

During World War II, the United States needed rubber to make many products used in the war, including truck tires and boots. So, the government challenged companies to help invent a synthetic rubber. James Wright, an engineer with General Electric Company, was one of the scientists trying to create this product. One day, Wright accidentally added boric acid to silicone oil as he experimented, which resulted in a thick, gooey product that had intriguing properties[1]. Wright tossed some on the floor and found that it bounced.

With some further experimentation, Wright discovered that this new material bounced and stretched even more than rubber. The material was named Nutty Putty. Wright kept working with the substance, but he could not figure out a good use for it. It certainly did not have the right properties to be used as a synthetic rubber. Soon, Nutty Putty was shared with other scientists in different labs. Still, no one could come up with a purpose for this odd material.

In addition to sharing Nutty Putty with other scientists, General Electric engineers passed it around at parties to let people explore the product. At one of these events, toy-store owner Ruth Fallgatter saw the material and realized the potential of this putty as a toy. She called the toy Bouncing Putty, and it became quite popular. Later on, Peter Hodgson bought the rights to the putty. He then packaged it inside a plastic egg and sold it as Silly Putty. Hodgson made millions of dollars on this toy putty before it was sold to the company that makes Crayola crayons.

Ever since its invention in 1943, people have been discovering new purposes for Silly Putty. It's been used for everything from lifting pencil marks or ink from paper to collecting lint found on fabric. The toy putty was even sent on the *Apollo 8* mission to the Moon where it was used to keep tools from floating around in space.

So, what exactly is Silly Putty made of? The classic coral-colored product is 65% dimethylsiloxane (boric acid with polymers), 17% silica, 9% Thixatrol ST (castor oil derivative), 4% polydimethylsiloxane, 1% decamethylcyclopentasiloxane, 1% glycerine, and 1% titanium dioxide.

Recognize, Recall, and Reflect

1. What was James Wright trying to do when he discovered the material that would become Silly Putty?

2. What is a non-Newtonian fluid?

3. Why was Nutty Putty considered interesting enough to make it into a toy?

---

1 Although James Wright usually gets the credit for the invention of this substance, there are also reports that it was first created by Earl Warrick. This scientist worked for the Dow Corning Company. Warrick has a patent for the material, dated December 2, 1942. Wright also has a patent, filed on February 13, 1951. Both patents are similar, but it is interesting to note that Warrick's patent was filed first.

Investigate

In this activity, you will explore properties of toy putty.

Materials

For each group of students:

- 1 sample of toy putty

- 1 hammer

- Indirectly vented chemical splash safety goggles (1 pair per student)

Safety Note: Do not ingest the putty. Wear indirectly vented chemical splash safety goggles during the setup, hands-on, and takedown segments of the activity. Use caution when using hand tools as they can cut or puncture skin. Wash your hands with soap and water immediately after completing this activity.

Create, Innovate, and Investigate

- Begin by observing the putty. What do you notice about the texture and shape of it?

- Bend the putty in different directions. Does it have any unusual properties?

- Roll the putty into a ball. Then bounce it on the table and on the floor. What do you observe?

- Gently hit the putty with a hammer. What do you notice about the material?

Questions for Reflection

1. What did you observe about the texture of the putty?

2. What is different about how the putty behaves compared to other materials such as clay?

3. What happens when you apply force with a hammer to the putty?

Make Your Own Putty

Follow these two recipes to create your own putty.

Safety Note: Do not consume the glue or borax solution. Wear indirectly vented chemical splash safety goggles and nitrile gloves during the setup, hands-on, and takedown segments of the activity. Wash your hands with soap and water immediately after completing this activity.

Recipe 1: Red Putty

MATERIALS

For each group of students:

- ¼ cup white school glue

- ¼ cup water

- ⅛ cup water in separate cup

- 1 tbsp. borax (often found in the grocery store with laundry soap)

- 2 plastic cups

- Mixing bowl

- 2 stirrers or spoons

- Red food coloring

- Indirectly vented chemical splash safety goggles (1 pair per student)

- Nonlatex apron (1 per student)

- Nitrile gloves (1 pair per student)

DIRECTIONS

- Mix the ⅛ cup of water and glue together in a plastic cup.

- In another cup, mix the ¼ cup of water and the borax together.

- In the mixing bowl, add the glue and water mixture, and then add the borax solution while stirring. Once it has thickened, you can pick the putty up and mix it by hand.

- Add two to three drops of the red food coloring and kneed the ball with your hands to mix the food coloring throughout.

Recipe 2: Blue Putty

MATERIALS

For each group of students:

- ½ cup white school glue

- ¼ cup liquid starch

- Mixing bowl

- Large spoon

- Blue food coloring

- Indirectly vented chemical splash safety goggles (1 pair per student)

- Nonlatex apron (1 per student)

- Nitrile gloves (1 pair per student)

DIRECTIONS

- Place the glue in a mixing bowl.

- Add the starch to the glue.

- Mix the ingredients together.

- Add two to three drops of the blue food coloring and kneed the ball with your hands to mix the food coloring throughout.

Compare the properties of the two types of homemade putty. Which one bounces higher? Which one becomes more like a solid when pressed down with a spoon?

Apply and Analyze

Putty: How Does It Work?

Toy putty is a non-Newtonian fluid and, as such, does not behave like a typical liquid or solid. Non-Newtonian fluids typically flow when they are under low stress or force but will break when under high stress or force. If you take a large ball of toy putty and punch it, the fluid will act like a solid and appear to stiffen to block your hand. But if you take your fingers and gently press them into the toy putty, they will slide in as if the material was a thick liquid. Non-Newtonian fluids do not have a constant value of viscosity, which means their thickness and resistance to flow changes depending on different factors. Viscosity also influences flow rates. (For instance, thick pudding, which has a higher viscosity, pours much more slowly compared to water, which has a lower viscosity.)

Toy putty is considered a suspension, which is defined as a heterogeneous mixture that contains solid particles that do not dissolve but rather float throughout the liquid medium. Some of the factors that influence the behaviors of non-Newtonian fluids include particle size, particle distribution, particle shape, the interactions between particles, and the type and rate of deformation (e.g., pressing a finger down into the material, causing it to "deform," or change shape).

Although it is fun to imagine using different fluids to make a new toy, when developing practical products of varying viscosities, it is critical to know if a material is Newtonian (like water) or non-Newtonian (like toy putty) because the difference in properties influences manufacturing, packaging, and the product's potential uses. For example, think about everyday fluids with different viscosities, such as yogurt, mayonnaise, honey, or jelly. What are the challenges in trying to pour these fluids from one container to another?

Uses for Toy Putty

Toy putty has been adopted for all sorts of purposes and applications. Explore some of the many uses of the material on this website: *http://escapeadulthood.com/blog/2012-12-03/17-surprisingly-practical-uses-for-silly-putty.html*. Then do additional online research into the various applications for the product.

One of the proposed uses of toy putty is as a beverage insulator. Insulators prevent the flow of thermal energy. Toy putty is composed of water, which is a good thermal insulator. So, you could wrap the material around a cold can to keep the drink chilled. Putty could also be added to machines to prevent squeaking. A small piece of putty placed between two metal parts can keep the metal from rattling. What are your thoughts on these different uses? Which of the uses do you think is most helpful to your everyday life? Based on what you've read above, answer the questions below.

1. Imagine you had a food product that behaved like toy putty. What would be the challenges of adding blueberries to your food product?

2. What are some applications of toy putty that are being explored? Which of these applications do you find most useful? Which are least useful?

Design Challenge

Engineering is the application of scientific understanding through creativity, imagination, problem solving, and the designing and building of new materials to address and solve problems in the real world. You will be asked to take the science you have learned in this case and design a process or product to address a real-world issue of your choosing.

Engineers use the engineering design process as steps to address a real-world problem (see Figure 11.1). You will now use this process as you come up with a new application for toy putty or a toy putty–like material. In this case, you are asking the question (Step 1) of how materials like toy putty could be used for new purposes. Drawing on your creativity, you will then brainstorm (Step 2) a specific new use for toy putty (or another non-Newtonian fluid) that solves a problem. Afterward, you will create a plan (Step 3) for your product. Although you will not actually produce your product, you will create a sketch and/or a model of it (Step 4). Then, you will work with your classmates to think about how you would test (Step 5) and refine (Step 6) your product.

1. Ask Questions

Read the background information below about innovative uses for toy putty:

EXAMPLE 1

A group of students at Case Western Reserve University in Cleveland, Ohio, thought of an interesting new use for toy putty: as a filling for potholes. The students had been trying to think of an everyday problem that could be solved with a simple material. They figured that using toy putty to fix potholes was a good solution due to the putty's properties as a non-Newtonian fluid. Without any pressure, the product would act like a liquid to fill the holes. However, if a car applied force by running over it, the product would act like a solid. You can go to the following link to read more about the students' idea, which won them first prize in a 2012 engineering contest: *www.sciencemag.org/news/2012/04/silly-putty-potholes*.

FIGURE 11.1

The Engineering Design Process

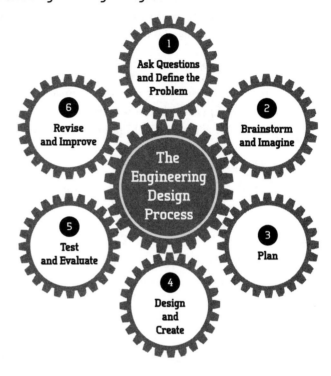

The Engineering Design Process

1. Ask Questions and Define the Problem
2. Brainstorm and Imagine
3. Plan
4. Design and Create
5. Test and Evaluate
6. Revise and Improve

EXAMPLE 2

Engineers have figured out that it is possible to make sports equipment with a toy putty–like fluid because it absorbs the energy of high-velocity objects. For example, U.S. and Canadian skiers in the 2006 Winter Olympics wore body armor that was made from a material similar to toy putty for protection. D3O, the company that made the armor, now makes a range of different products that feature soft, flexible material that can absorb high shock. Check out some of their products here: *www.d3o.com/faq*.

Now that you've read these examples, consider these questions: What new products could you make with a material like toy putty? What other products could be made that absorb impact? Are there other problems that toy putty can solve?

2. Brainstorm and Imagine

Brainstorm a product that incorporates toy putty or another non-Newtonian fluid to solve a problem. (For instance, this material's shock-absorbing properties might make it ideal for creating knee pads for skateboarding or roller skating. Perhaps the material could also be used to make bicycle helmets or bulletproof vests.)

3. Create a Plan

Create a plan for your product. Consider: (1) What is the purpose of your new product? (2) What are the benefits of the product? (3) What are the limitations of the product? Use the Product Planning Graphic Organizer (p. 176) to help you.

4. Design and Create

Consider these questions and considerations for your toy putty/non-Newtonian fluid product and its design.

- How would the product work?

- How would you overcome any limitations or drawbacks using toy putty/ non-Newtonian fluid?

- What technologies might need to be developed to create or manufacture this design?

- What are any constraints or drawbacks you can foresee with implementing this design?

- Would there be any safety concerns regarding your toy putty–based product?

Now, create a sketch of your product. Make sure your design incorporates the research and exploration you've done.

5. Test and Evaluate

Working with your classmates, come up with a way to test your design to see its effectiveness.

6. Revise and Improve

Give your plans to one of your classmates for review. Listen to his or her feedback on your design. What are some ways you can use the input to refine your design? Take some time to revise and make improvements.

Reflect

1. What technologies might need to be developed to create or manufacture this design?

2. What are any constraints or drawbacks you can foresee with implementing this design?

3. Would there be any environmental or human health concerns to using the product in this way?

Product Planning Graphic Organizer

Proposed Product Idea	
Pros (Benefits)	**Cons (Limitations)**

PUTTY IN YOUR HANDS

THE UNINTENDED DISCOVERY OF SILLY PUTTY

A Case Study Using the Discovery Engineering Process

Lesson Overview

In this lesson, students explore Silly Putty and non-Newtonian fluids. Silly Putty was an accidental discovery that has led to a number of new applications and products.

Lesson Objectives

By the end of this case study, students will be able to

- Describe the properties of a non-Newtonian fluid.

- Analyze the behavior of toy putty.

- Design a new product or application that uses toy putty or another non-Newtonian fluid.

The Case Study Approach

This lesson uses a case study approach. Explaining the purpose of case studies will encourage your students to relate to the material and engage with the problem. At the heart of each case study in this book is a true story, one that describes how someone in his or her everyday life or during a routine workday made an observation or did a simple experiment that led to a new insight or discovery. Case studies are designed to get students actively engaged in the process of problem solving. The narrative of the case supplies authentic details that place the student in the role of the inventor and provide scaffolds for critical thinking and deep reflection. A case is more than a paragraph to read or a story to analyze but rather a way of framing problems, synthesizing what is known, and thinking creatively about new applications and solutions. In this lesson, students consider how Silly Putty was discovered and work together to think about new applications for toy putty to solve real-life problems.

Use of the Case

Due to the nature of these case studies, teachers may elect to use any section in each case for their instructional needs. The sections are sequenced in order (scaffolded) so that students think more deeply about the science involved in the case and develop understanding of engineering in the context of science.

Curriculum Connections

Lesson Integration

You could use this case as a way to integrate engineering into a discussion of states of matter or into a lesson on atomic structure or chemistry. This case can be followed with investigations of non-Newtonian fluids. Here are some resources on non-Newtonian fluids that you can share with your class:

- This video provides a fun demonstration of walking and moving through a non-Newtonian fluid: *www.youtube.com/watch?v=RIUEZ3AhrVE*.

- This video explores the strength of cornstarch and water: *www.youtube.com/watch?v=Sl0BHueSjvA*. It can be followed by an explanation of non-Newtonian fluids: *www.sciencelearn.org.nz/resources/1502-non-newtonian-fluids*.

There are a number of other resources and videos about non-Newtonian fluids available online that you may want to explore before using the case.

Related Next Generation Science Standards
PERFORMANCE EXPECTATIONS

- MS-PS1-1. Develop models to describe the atomic composition of simple molecules and extended structures. (Substances are made of different types of atoms, which combine with one another in various ways.)

- MS-PS1-4. Develop a model that predicts and describes changes in particle motion, temperature, and state of a pure substance when thermal energy is added or removed.

- HS-PS2-6. Communicate scientific and technical information about why the molecular-level structure is important in the functioning of designed materials.

- HS-ETS1-2. Design a solution to a complex real-world problem by breaking it down into smaller, more manageable problems that can be solved through engineering.

- HS-ETS1-3. Evaluate a solution to a complex real-world problem based on prioritized criteria and trade-offs that account for a range of constraints, including cost, safety, reliability, and aesthetics, as well as possible social, cultural, and environmental impacts.

SCIENCE AND ENGINEERING PRACTICES

- Analyzing and Interpreting Data
- Engaging in Argument From Evidence
- Constructing Explanations and Designing Solutions

CROSSCUTTING CONCEPT

- Structure and Function

Related National Academy of Engineering Grand Challenge

- Engineer the Tools of Scientific Discovery

Lesson Preparation

You will need to make copies of the entire student section for the class. Students will need internet access at various points in the lesson. Alternatively, you can project videos or print and distribute copies of online content for the class. The materials for the Make Your Own Putty activity in the Investigate section are messy, so newsprint or plastic sheeting placed on tables can help facilitate cleanup. Look at the Teaching Organizer (Table 11.1, p. 181) for suggestions on how to organize the lesson. It may be useful to review the properties of solids, liquids, and gases.

Materials (Main Investigation)

For each group:

- 1 sample of putty
- 1 hammer
- Indirectly vented chemical splash safety goggles (1 pair per student)

Safety Note for Students: Do not ingest the putty. Wear indirectly vented chemical splash safety goggles during the setup, hands-on, and takedown segments of the activity. Use caution when using hand tools as they can cut or puncture skin. Wash your hands with soap and water immediately after completing this activity.

Materials (Make Your Own Putty)

Groups of students will need the following materials for each recipe:

RECIPE 1: RED PUTTY

- ¼ cup white school glue
- ¼ cup water
- ⅛ cup water in separate cup
- 1 tbsp. borax (often found in the grocery store with laundry soap)
- 2 plastic cups
- Mixing bowl
- 2 stirrers or spoons
- Red food coloring
- Indirectly vented chemical splash safety goggles (1 pair per student)
- Nonlatex apron (1 per student)
- Nitrile gloves (1 pair per student)

RECIPE 2: BLUE PUTTY

- ½ cup white school glue
- ¼ cup liquid starch
- Mixing bowl
- Large spoon
- Blue food coloring
- Indirectly vented chemical splash safety goggles (1 pair per student)
- Nonlatex apron (1 per student)
- Nitrile gloves (1 pair per student)

Safety Note for Students: Do not consume the glue or borax solution. Wear indirectly vented chemical splash safety goggles and nitrile gloves during the setup, hands-on, and takedown segments of the activity. Wash your hands with soap and water immediately after completing this activity.

Time Needed

100 minutes

TABLE 11.1

Teaching Organizer

Section	Time Suggested	Materials Needed	Additional Considerations
The Case	5 minutes	Student packet	Could be read in class or as a homework assignment prior to class
Investigate	40 minutes	Student packet, 1 sample of putty, 1 hammer, indirectly vented chemical splash safety goggles, nonlatex apron, Nitrile gloves, Recipe 1 materials, Recipe 2 materials	The main activity can be done in small groups. For the Make Your Own Putty activity, small groups of students can either make both recipes, or some groups can make Recipe 1 and others can make Recipe 2. Then the groups can share their putty creations.
Apply and Analyze	10 minutes	Student packet, internet access	Whole-class or individual activity
Design Challenge	45 minutes	Student packet, internet access	Small-group activity

Teacher Background Information

Students often have questions about non-Newtonian fluids and why they behave differently from most other fluids. You may want to show them a video about how non-Newtonian fluids behave. See, for example, *www.youtube.com/ watch?v=G1Op_1yG6lQ*.

Vocabulary

- heterogeneous
- mixture
- non-Newtonian fluid
- states of matter
- suspension
- viscosity

Teacher Answer Key

Recognize, Recall, and Reflect

1. **What was James Wright trying to do when he discovered the material that would become Silly Putty?**

    He was trying to create a new synthetic rubber that could be used in products needed in World War II, like truck tires and boots.

2. **What is a non-Newtonian fluid?**

    Non-Newtonian fluids typically flow like a liquid when under low stress or force but behave like a solid when under high stress or force.

3. **Why was Nutty Putty considered interesting enough to make it into a toy?**

    Nutty Putty bounced and stretched in unusual ways.

Questions for Reflection

1. **What did you observe about the texture of the putty?**

    It has a smooth texture. The material acts like a liquid or a solid depending on how much pressure is applied to it.

2. **What is different about how the putty behaves compared to other materials such as clay?**

    It exhibits properties of both solids and liquids.

3. **What happens when you apply force with a hammer to the putty?**

    It will break like a solid when force is applied.

Apply and Analyze

1. **Imagine you had a food product that behaved like toy putty. What would be the challenges of adding blueberries to your food product?**

    Answers will vary, but one problem is getting berries to mix into the product evenly without breaking.

2. **What are some applications of toy putty that are being explored? Which of these applications do you find most useful? Which are least useful?**

Students will find various examples of applications in their online research. Answers may include that the putty can be used as a pencil grip, to open a jar, or as a ball. Answers about which products are most/least useful will vary.

Reflect

1. **What technologies might need to be developed to create or manufacture this design?**

   Answers will vary.

2. **What are any constraints or drawbacks you can foresee with implementing this design?**

   Answers will vary.

3. **Would there be any environmental or human health concerns to using the product in this way?**

   Answers will vary.

Assessment

The Design Challenge can be assessed using the rubric in the appendix (p. 377). This case also provides a rich opportunity to examine students' concepts of states of matter. For example, you might ask students to draw a picture of the molecular structure of a solid, liquid, and gas. Then ask them to draw a picture of the molecular structure of a non-Newtonian fluid. Having students draw what they think is happening at the molecular level allows you to examine their concepts and provide scaffolding if needed.

Extensions

This lesson can be followed with lessons about states of matter and other non-Newtonian fluids.

Resources and References

The Backyard Scientist. 2016. "How strong is Oobleck?" YouTube video. *www.youtube.com/watch?v=Sl0BHueSjvA.*

Crayola. Silly Putty history. *https://web.archive.org/web/20080603053016/http://www.crayola.com:80/mediacenter/index.cfm?display=press_release&news_id=164* (accessed December 14, 2017).

Cuda Kroen, G. 2012. Silly Putty for potholes. *Science. www.sciencemag.org/news/2012/04/silly-putty-potholes.*

General Electric Company. 2009. Non-Newtonian fluid. *www.ge.com/press/scienceworkshop/docs/pdf/Non_Newtonian_Fluid_with_Standards.pdf.*

Hiskey, D. 2011. Silly Putty was invented by accident. Today I Found Out. *www.todayifoundout.com/index.php/2011/11/silly-putty-was-invented-by-accident.*

How Stuff Works. How to get Silly Putty out of clothes. *https://home.howstuffworks.com/how-to-get-silly-putty-out-of-clothes.htm* (accessed December 14, 2017).

Instructables. How to: Make non-Newtonian fluid (& experiment with it). *www.instructables.com/id/How-To%3A-Make-Non-Newtonian-Fluid-%26-Experiment-wit* (accessed December 14, 2017).

Kids Discover. 2013. Weird science: The invention of Silly Putty. *www.kidsdiscover.com/quick-reads/weird-science-the-accidental-invention-of-silly-putty* (accessed December 14, 2017).

Kotecki, J. 2012. 17 surprisingly practical uses for Silly Putty. Escape Adulthood. *http://escapeadulthood.com/blog/2012-12-03/17-surprisingly-practical-uses-for-silly-putty.html.*

Lamar University. 2014. "Fun with non-Newtonian fluid-Lamar University." YouTube video. *www.youtube.com/watch?v=RIUEZ3AhrVE.*

Oullette, J. 2016. Has the mystery behind this non-Newtonian fluid been solved at last? Gizmodo. *https://gizmodo.com/has-the-mystery-behind-this-non-newtonian-fluid-been-so-1775600921.*

Rohrig, B. 2017. No-hit wonder! *ChemMatters. www.acs.org/content/dam/acsorg/education/resources/highschool/chemmatters/issues/2016-2017/February%202017/chemmatters-feb2017-d3o.pdf.*

Science Learning Hub. 2012. Non-Newtonian fluids. *www.sciencelearn.org.nz/resources/1502-non-newtonian-fluids.*

The Slow Down Show. 2013. "Non-Newtonian liquid in slow motion!" YouTube video. *www.youtube.com/watch?v=G1Op_1yG6lQ.*

Zarda, B. 2009. The incredibly wide world of smart material d3o. *Popular Science. www.popsci.com/gear-amp-gadgets/article/2009-08/incredibly-wide-world-smart-material-d3o.*

SWEET AND SOUR

The Rise and Fall ... and Rise ... of Saccharin

A Case Study Using the Discovery
Engineering Process

Introduction

Have you ever chewed sugar-
free gum or drank a calorie-free
soda and wondered, "How do
they sweeten these foods without
sugar?" The answer to the question may very well be an artificial sweetener called saccharin
($C_7H_5NO_3S$). Saccharin is approximately 300 to 400 times sweeter
tasting than sugar. It is water-

soluble and non-nutritive (having very few to no calories). Artificial sweeteners
appear in many processed foods like gum, candy, canned fruit, desserts, baked
goods, and tabletop sweeteners.

Developed in the late 19th century, saccharin found its fame during sugar ration-
ing in World War I. It has been welcomed by diabetics and dieters alike as a sugar
substitute because it is indigestible by the human digestive tract. That is, it does not
require processing by the pancreas or add inches to the waistline when consumed.

Studies done on saccharin in the late 20th century suggested that the sweetener
caused cancer in lab rats. Later research discredited these studies and concluded

that saccharin does not pose a cancer risk to humans. Today, it is the third most popular artificial sweetener in the United States.

Lesson Objectives

By the end of this case study, you will be able to

- Describe the discovery that led to the development of saccharin.

- Explore how redox reactions change the chemical orientation/composition of compounds.

- Explain the dangers of poorly designed or interpreted studies.

The Case

Read the passage below about an accidental discovery of saccharin.

Dr. Constantin Fahlberg knew a lot about sugar. In fact, he not only researched sugar, but he also worked for a sugar company where he tested the purity of imported cane sugar shipments. One of his coworkers was Dr. Ira Remsen, a chemist who studied sulfobenzoic compounds, which are compounds formed when a sulfur-containing molecule is added to a molecule of benzoic acid. These chemicals are typically used as a food preservative due to their antimicrobial properties. Fahlberg was invited to conduct his sugar research alongside Remsen in his laboratory.

In the summer of 1878, Fahlberg sat down to eat his dinner after a long day of work in the lab. He bit into his dinner roll and noticed it tasted exceptionally sweet, far sweeter than a dinner roll should be. He realized that he had not washed his hands before eating, and the source of the sweetness was likely due to a chemical he had accidentally spilled on himself at work. He immediately rushed back to the lab to search for the source of the sweetness. Since he did not know which beaker contained the secretly sweet chemical, he tasted all the beakers at his workbench! He eventually found the source, which was a beaker containing sulfobenzoic acid, phosphorus chloride, and ammonia. Earlier that day, the components in the beaker had been boiled, which created benzoic sulfimide. Fahlberg realized that he had made that compound before but never had any reason to taste it. His accidental taste test led to the development of the first artificial sweetener, which was named saccharin.

Recognize, Recall, and Reflect

1. How did Dr. Fahlberg accidentally discover saccharin?

2. What three ingredients created benzoic sulfimide (saccharin)?

3. What caused these ingredients to produce the benzoic sulfimide?

Investigate

In this section, you will learn about a process similar to the one that creates saccharin. You may have noticed that some fences or cars will grow a red-colored crust when exposed to rain and wind over time. The crust is rusted iron. Under normal conditions, iron is a very strong metal. Yet rusted iron is weak and brittle. This is because the iron has undergone a chemical reaction. Oxygen in the air and water take electrons away from atoms of iron. This causes a permanent chemical change (making iron oxide, or rust) through a process called an oxidation-reduction or redox reaction.

Redox reactions involve the transfer of electrons, through oxidation and reduction. An atom's oxidation number (also called oxidation state) represents the number of electrons lost or gained by an atom in a compound. Oxidation refers to the loss of electrons from an atom or compound, resulting in an increased oxidation number and an increase in positive charge. Conversely, reduction refers to an atom or compound gaining electrons, resulting in a lower or reduced oxidation number and a more negative charge. Metallic elements like iron, sodium, and magnesium have a greater tendency to oxidize than other elements. They are called electron donors and have low electronegativity values—that is, they do not tend to attract shared electrons to themselves within molecules. Nonmetal elements such as nitrogen, chlorine, and oxygen have a tendency to gain electrons. These elements are called electron acceptors and are more electronegative than metals. They tend to attract shared electrons (which have a negative charge) within molecules.

This tendency for atoms to gain or lose electrons is due to the stability of electrons in their outermost (valence) shell. The noble gases (Group 8 on the periodic table) are difficult to ionize, and they rarely undergo chemical reactions because their electron configurations are extremely stable. This is because their valence shell is completely filled. Metals and nonmetals tend to undergo oxidation or reduction until their valence shell has the same electron configuration as the nearest noble gas on the periodic table. The metals sodium and magnesium, for example, both undergo oxidation until their valence shell resembles that of neon. Carbon and fluorine, two nonmetals, undergo reduction until they resemble neon, as well. With the exception of helium, whose valence shell holds two electrons, all other noble gases have eight valence electrons, which is why the octet rule is used as a guideline for the number of valence electrons an atom is likely to have in a molecule.

In a redox reaction, one element is always being oxidized (giving up electrons) and another is being reduced (gaining electrons), which means oxidation and reduction occur simultaneously as electrons are transferred from one element to the other. It's helpful to understand redox reactions by identifying the elements that are oxidized and reduced and labeling their respective oxidation states.

As an example, let's consider the following chemical equation: $2 H_2 + O_2 \rightarrow 2 H_2O$. In the synthesis of water, hydrogen atoms are being oxidized and oxygen atoms are being reduced. Hydrogen goes from an oxidation state of 0 to +1, because each hydrogen atom loses an electron. The plus indicates there is more positive charge (protons) than negative charge (electrons). Oxygen has changed from a 0 to –2 oxidation state, reflecting the two additional negative charges (electrons), one gained from each of the two hydrogen atoms.

The following equations are common redox reactions. First, identify the oxidizing and reducing elements. Then, describe the oxidation states.

- Rusting of iron: $4\,Fe + 3\,O_2 \rightarrow 2\,Fe_2O_3$

- Acetylene combustion: $2\,C_2H_2 + 5\,O_2 \rightarrow 4\,CO_2 + 2\,H_2O$

- Cellular respiration of glucose: $C_6H_{12}O_6 + 6\,O_2 \rightarrow 6\,CO_2 + 6\,H_2O$

Now that you have read the above background information, complete the following activity on redox reactions. You will experiment with redox reactions to explore how the transfer of electrons changes the physical and chemical properties of elements.

Materials

For each group of students:

- Metallic zinc strip and a metallic copper strip

- 50 ml beakers (2 per group)

- 3 ml of 0.1 molar copper (II) sulfate and 3 ml of 0.1 molar silver nitrate

- Indirectly vented chemical splash safety goggles (1 pair per student)

- Nonlatex apron (1 per student)

- Nitrile gloves (1 pair per student)

Safety Note: Wear indirectly vented chemical splash safety goggles, a nonlatex apron, and nitrile gloves during the setup, hands-on, and takedown segments of the activity. Follow your teacher's instructions for disposing of waste materials. If you get a chemical in your eye, use an eyewash station immediately. Use caution in working with silver nitrate—it will stain the skin. Wash your hands with soap and water immediately after completing this activity.

Create, Innovate, and Investigate

- Make observations of both the zinc and copper metals. Note their physical properties.

- Prepare the beakers for the redox reactions:

- Reaction I: Add the 3 ml of copper (II) sulfate to the first beaker.

- Reaction II: Add the 3 ml of silver nitrate to the second beaker.

- Add the strip of zinc to the copper (II) sulfate beaker. Observe any physical changes.

- Add the strip of copper to the silver nitrate beaker. Observe any physical changes.

Questions for Reflection

1. Which metals were more reactive? Zinc or copper? Copper or silver? Why?

2. What are the chemical equations for Reactions I and II?

3. Which elements were oxidized and reduced in Reaction I? Reaction II?

Apply and Analyze

The Process to Make Saccharin

The original process to make saccharin began with the chemical toluene. Using the Remsen-Fahlberg synthesis, toluene ($C_6H_5CH_3$) reacts with chlorosulfonic acid ($ClSO_3H$) to create two structural isomers—chemicals with the same atoms arranged in different ways. One of the structural isomers is discarded. The other structural isomer is converted to sulfonamide with ammonia (NH_3). It is then oxidized to create saccharin. Learn more about toluene here: *www.atsdr.cdc.gov/phs/phs.asp?id=159&tid=29*. Then look at Figure 12.1, which shows the molecular formula for saccharin.

FIGURE 12.1

Molecular Formula for Saccharin

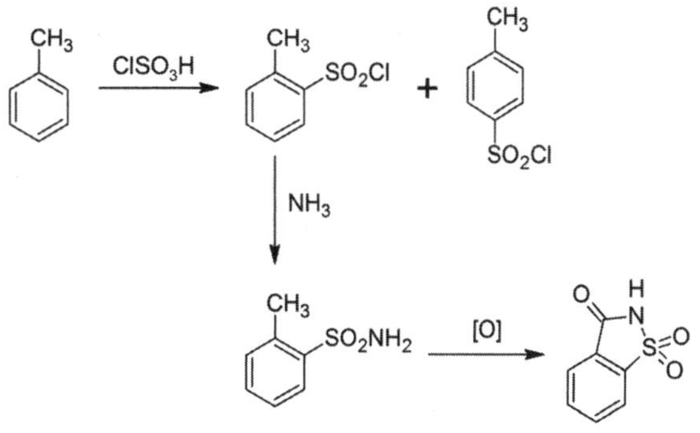

The Safety of the Sweetener

From 1970 to 2000, saccharin was the focus of studies indicating a link between the artificial sweetener and bladder cancer in laboratory rodents. However, recent research has shown those studies had methodological flaws and suggests that saccharin is relatively safe for human consumption. In 2010, the Environmental Protection Agency officially stated that it no longer considers saccharin a potential hazard to human health.

Among other things, a good scientific study is objective and unbiased, includes validated and appropriate ways of collecting data, and passes an extensive peer review. Sometimes studies, like the early ones linking saccharin to cancer, are poorly executed or draw incomplete or incorrect conclusions. In these cases, scientists strive to uncover accurate results by doing follow-up research that is more thorough. Look at the topics below. These are cases in which initial studies drew incomplete or incorrect conclusions that were later debunked by additional research. Pick one of the topics, look over the study or studies that provided the incomplete or incorrect results, and then research how these results were debunked. (You may also find other examples of such studies online or through your school library.)

- Saccharin and bladder cancer: *www.ncbi.nlm.nih.gov/pmc/articles/PMC1637197*

- Vaccines and autism: *http://lbihealth.com/wp-content/uploads/2018/03/vaccines-autism.pdf*

- Benefits of alcohol and heart disease: *www.annalsofepidemiology.org/article/S1047-2797(07)00007-5/pdf*

- SSRIs (antidepressant medications) and adolescent suicide: *http://ajp.psychiatryonline.org/doi/pdfplus/10.1176/appi.ajp.160.4.790*

After researching your topic, answer the questions below:

1. What are the central issues in the topic you chose to focus on?

2. How were previous studies into this topic flawed or misinterpreted?

Design Challenge

Design a campaign to communicate to the public the latest research on the topic you picked in the Apply and Analyze section. (For example, your campaign can focus on informing the public that no link exists between vaccinations and autism.) Write out a plan for your campaign, describing how you will disseminate information to the public. (You can come up with public service announcements, an app that communicates information, and more.)

Remember that the engineering design process (Figure 12.2) contributes to the development of products such as new vaccines, antidepressants, and artificial sweeteners. What steps in the engineering design process would you highlight to ease public concern about using new products that may have gotten a bad reputation from inaccurate scientific studies? Include this in your campaign. (For instance, if you are designing an educational campaign about vaccines, you can emphasize that new vaccines are thoroughly tested and evaluated in Step 5 of the engineering design process. This could give the public more confidence in the product.)

FIGURE 12.2

The Engineering Design Process

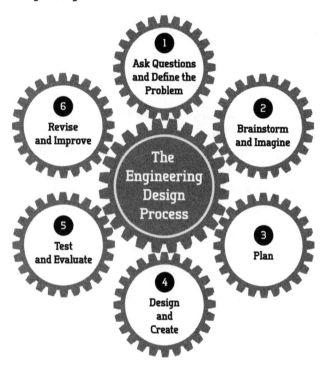

Reflect

1. How will your campaign communicate to the public the latest scientific research on your topic?

2. What strategies could you employ to assuage public fears about this topic?

3. What are any constraints or drawbacks you can foresee with implementing your campaign? How can you address these?

SWEET AND SOUR

THE RISE AND FALL ... AND RISE ... OF SACCHARIN

A Case Study Using the Discovery Engineering Process

Lesson Overview

In this lesson, students learn about the accidental discovery of saccharin and the chemistry behind this innovation. They learn about redox reactions and how to identify the oxidizing and reducing elements. Students use this information to research a substance that was banned based upon incorrect evidence and present the scientific evidence to the public through a public campaign. This lesson is best suited for high school students with some knowledge of redox reactions.

Lesson Objectives

By the end of this case study, students will be able to

- Describe the discovery that led to the development of saccharin.

- Explore how redox reactions change the chemical orientation/composition of compounds.

- Explain the dangers of poorly designed or interpreted studies.

The Case Study Approach

This lesson uses a case study approach. Explaining the purpose of case studies will encourage your students to relate to the material and engage with the problem. At the heart of each case study in this book is a true story, one that describes how someone in his or her everyday life or during a routine workday made an observation or did a simple experiment that led to a new insight or discovery. Case studies are designed to get students actively engaged in the process of problem solving. The narrative of the case supplies authentic details that place the student in the role of the inventor and provide scaffolds for critical thinking and deep reflection. A case is more than a paragraph to read or a story to analyze but rather a way of framing problems, synthesizing what is known, and thinking creatively about new applications and solutions. In this lesson, students consider how saccharin was discovered, learn about studies on saccharin that had methodological flaws, and think of ways to correct public knowledge about a topic after misinformation has spread.

Use of the Case

Due to the nature of these case studies, teachers may elect to use any section in each case for their instructional needs. The sections are sequenced in order (scaffolded) so students think more deeply about the science involved in the case and develop an understanding of engineering in the context of science.

Curriculum Connections

Lesson Integration

This lesson may be taught during a unit on single replacement chemical reactions. It also fits well into a lesson on redox reactions or balancing chemical equations. This lesson requires students to have some understanding of atomic bonding and the periodic table.

Related Next Generation Science Standards
PERFORMANCE EXPECTATIONS

- HS-PS2-6. Communicate scientific and technical information about why the molecular-level structure is important in the functioning of designed materials.

- HS-ETS1-2. Design a solution to a complex real-world problem by breaking it down into smaller, more manageable problems that can be solved through engineering.

- HS-ETS1-3. Evaluate a solution to a complex real-world problem based on prioritized criteria and trade-offs that account for a range of constraints, including cost, safety, reliability, and aesthetics, as well as possible social, cultural, and environmental impacts.

SCIENCE AND ENGINEERING PRACTICES

- Analyzing and Intepreting Data
- Engaging in Arguing From Evidence
- Constructing Explanations and Designing Solutions

CROSSCUTTING CONCEPT

- Structure and Function

Related National Academy of Engineering Grand Challenge

- Engineer the Tools of Scientific Discovery

Lesson Preparation

You will need to make copies of the entire student section for the class. Students will need internet access at various points in the lesson. Alternatively, you can project videos or print and distribute copies of online content. Before beginning, you may wish to review how to use the periodic table to determine electronegativity of elements. For the Investigate section, have solutions of 0.1 molar silver nitrate and 0.1 molar copper (II) sulfate prepared. You may wish to place these into 3 ml droppers. Also, cut small zinc and copper strips for each group (approximately 1 inch in length). The strips should be short enough to be covered by 3 ml of liquid. If you do not have the materials available for this activity, you may wish to use an online simulator. Look at the Teaching Organizer (Table 12.1) for suggestions on how to organize the lesson.

Materials

For each group of students:

- Metallic zinc strip and metallic copper strip
- 50 ml beakers (2 per group)
- 3 ml of 0.1 molar copper (II) sulfate and 3 ml of 0.1 molar silver nitrate
- Indirectly vented chemical splash safety goggles (1 pair per student)
- Nonlatex apron (1 per student)
- Nitrile gloves (1 pair per student)

Safety Note for Students: Wear indirectly vented chemical splash safety goggles, a nonlatex apron, and nitrile gloves during the setup, hands-on, and takedown segments of the activity. Follow your teacher's instructions for disposing of waste materials. If you get a chemical in your eye, use an eyewash station immediately. Use caution in working with silver nitrate—it will stain the skin. Wash your hands with soap and water immediately after completing this activity.

Time Needed

105 minutes

TABLE 12.1

Teaching Organizer

Section	Time Suggested	Materials Needed	Additional Considerations
The Case	5 minutes	Student packet	Could be read in class or as a homework assignment prior to class
Investigate	25 minutes	Student packet, metallic zinc strip and metallic copper strip, 50 ml beakers (2 per group), 3 ml of 0.1 molar copper (II) sulfate and 3 ml of 0.1 molar silver nitrate, indirectly vented chemical splash safety goggles, nonlatex apron, nitrile gloves	Recommended as a small-group activity
Apply and Analyze	30 minutes	Student packet, internet access	Small-group activity or done in pairs
Design Challenge	45 minutes	Student packet, internet access	Small-group activity or done in pairs

Vocabulary

- ammonia
- antimicrobial
- artificial sweetener
- benzoic sulfimide
- electronegativity
- isomer
- metallic elements
- nonmetallic elements
- non-nutritive
- octet rule

- oxidation
- phosphorus chloride
- redox
- reduction
- saccharin
- sucrose
- sulfobenzoic acid
- sulfobenzoic compounds
- valence
- water-soluble

Teacher Answer Key

Recognize, Recall, and Reflect

1. **How did Fahlberg accidentally discover saccharin?**

   He was eating dinner and his roll tasted extra sweet. He realized he had not washed his hands after working in his lab and that one of the chemicals must have gotten on the roll. So, he searched his lab for the source of the sweetener.

2. **What three ingredients created benzoic sulfimide (saccharin)?**

   Sulfobenzoic acid, phosphorus chloride, and ammonia

3. **What caused these ingredients to produce the benzoic sulfimide?**

   They were boiled together.

Questions for Reflection

1. **Which metals were more reactive? Zinc or copper? Copper or silver? Why?**

   Zinc is more reactive than copper because it displaces the copper ions in the solution. Similarly, copper is more reactive than silver because it displaces the silver ions in the solution.

2. **What are the chemical equations for Reactions I and II?**

   Reaction I: $CuSO_4 (aq) + Zn (s) \rightarrow ZnSO_4 (aq) + Cu (s)$
   Reaction II: $Cu (s) + 2AgNO_3 (aq) \rightarrow Cu(NO_3)_2 (aq) + 2Ag (s)$

3. **Which elements were oxidized and reduced in Reaction I? Reaction II?**

   Reaction I: oxidized: zinc; reduced: copper
   Reaction II: oxidized: copper; reduced: silver

Apply and Analyze

1. **What are the central issues in the topic you chose to focus on?**

   Student responses will vary according to their study.

2. **How were previous studies into this topic flawed or misinterpreted?**

   Examples include but are not limited to small sample size, poor methodology, no control group, selection bias (nonrandomized sample), replication bias (unable to reproduce results), cannot compare across studies due to varying protocols, no peer review, only one to two studies, and conflict of interest.

Reflect

1. **How will your campaign communicate to the public the latest scientific research on your topic?**

   Answers will vary.

2. **What strategies could you employ to assuage public fears about this topic?**

   Answers will vary.

3. **What are any constraints or drawbacks you can foresee with implementing your campaign? How can you address these?**

   Answers will vary.

Assessment

You can evaluate the students' descriptions of their public information campaigns in the Design Challenge to assess their understanding of the studies they chose to focus on, how some accepted claims get debunked, and the engineering design process.

Extensions

This case can be followed with lessons on organic chemistry, isomers, and substitution reactions.

Resources and References

Alama, A., C. Bruzzo, Z. Cavalieri, and A. Forlani. 2011. Inhibition of the nicotinic acetylcholine receptors by cobra venom α-neurotoxins: Is there a perspective in lung cancer treatment? *PlosOne* 6 (6): e20695. *www.ncbi.nlm.nih.gov/pmc/articles/PMC3113800*.

ATSDR. 2015. Public health statement for toluene. CDC. *www.atsdr.cdc.gov/phs/phs.asp?id=159&tid=29*.

Fillmore, K. et al. 2007. Moderate alcohol use and reduced mortality risk: Systematic error in prospective studies and new hypotheses. *Annals of Epidemiology* 17 (5s): S16–S23. *www.annalsofepidemiology.org/article/S1047-2797(07)00007-5/pdf*.

Hicks, J. 2010. The pursuit of sweet. Science History Institute. *www.sciencehistory.org/distillations/magazine/the-pursuit-of-sweet*.

Hodgin, G. The history, synthesis, metabolism and uses of artificial sweeteners. *http://monsanto.unveiled.info/products/aspartme.htm*.

Khan, A., S. Khan, R. Kolts, and W. Brown. 2003. Suicide rates in clinical trials of SSRIs, other antidepressants, and placebo: Analysis of FDA reports. *American Journal of Psychiatry* 160: 790–792. *https://ajp.psychiatryonline.org/doi/pdfplus/10.1176/appi.ajp.160.4.790.*

National Cancer Institute. 2016. Artificial sweeteners and cancer. *www.cancer.gov/about-cancer/causes-prevention/risk/diet/artificial-sweeteners-fact-sheet.*

Offit, P. 2005. Vaccines and autism. Immunization Action Coalition. *http://lbihealth.com/wp-content/uploads/2018/03/vaccines-autism.pdf.*

Reuber, M. D. 1978. Carcinogenicity of saccharin. *Environmental Health Perspectives* 25 (Aug.): 173–200. *www.ncbi.nlm.nih.gov/pmc/articles/PMC1637197.*

Smallwood, K. 2014. The accidental discovery of saccharin, and the truth about whether saccharin is bad for you. Today I Found Out. *www.todayifoundout.com/index.php/2014/05/saccharin-discovered-accident.*

SAVING LIVES THROUGH AN ACCIDENT

Safety Glass

A Case Study Using the Discovery Engineering Process

Introduction

Hard, sharp, and smooth—glass is an extraordinary material with many practical uses. During the Stone Age, humans created tools with obsidian, a naturally occurring volcanic glass. Evidence of glass-making technologies can be traced as far back as around 3500 BC to civilizations in Egypt and Mesopotamia. Since that time, many inventors have experimented with glass-making technologies and have improved on one another's techniques. An accidental discovery in a chemistry lab in 1903 helped make broken glass less hazardous. This innovation has had profound implications for the use of glass in modern products and has created exciting new applications for a material that humans have been using for millennia!

Lesson Objectives

By the end of this case study, you will be able to

- Describe the characteristics of two types of safety glass (laminated and tempered).

- Analyze the benefits and limitations of two types of safety glass as a material for consumer products.

- Design a new application for safety glass that solves a problem.

The Case

Read the following summary of information from the articles "The Discovery of Safety Glass," published in *A Flash of Genius* by Alfred Garrett, and "Safety Glass," published in an *Introduction to Polymer Chemistry* by Charles Carraher. Once you are finished reading, answer the questions that follow.

Edouard Benedictus was a chemist living in France at the start of the 20th century. In 1903, Benedictus was rearranging some flasks on a high shelf in his laboratory when one slipped from his grasp and fell to the floor from a height of at least 11 feet. Although Benedictus heard the flask shatter, he was surprised to see that it had not broken apart on impact with the floor! Large cracks ran through the flask, but it held its shape. The glass fragments appeared to stick together for the most part, and not a single sharp shard had broken free. What was going on? Looking closer at the cracked flask, Benedictus realized there was a clear, thin film coating the inside that held the broken pieces in place. He spoke with his assistant and learned that the flask had held a solution of cellulose nitrate (a compound that is used in materials such as paints, plastics, and gunpowder). Benedictus and his assistant deduced that the solution of cellulose nitrate had dried, leaving behind a clear layer on the inside of the flask. The assistant hadn't notice the transparent layer and had returned the glassware to the shelf with the rest of the clean flasks.

A use for this new material wasn't immediately obvious to Benedictus and his assistant. But a short time later, Benedictus learned from a newspaper article about a young girl who had been badly cut by sharp pieces of flying glass during an automobile accident. Such injuries from glass shards were not uncommon during this period in history. After hearing about another similar accident, Benedictus remembered how the broken glass pieces from his dropped flask had stayed together instead of flying apart. This gave Benedictus an idea. He experimented with placing the cellulose nitrate between two flat pieces of glass and used pressure to bond these glass pieces together. Since cellulose nitrate dries clear, the final glass material was still transparent, which made it very useful. In 1909, Benedictus patented his new product—a layer of cellulose nitrate sandwiched between two layers of glass.

In 1919, Ford Motor Company began using this type of safety glass, referred to as "laminated glass," for the windshields of its cars. (See Figure 13.1.) This reduced the number of injuries and deaths caused by sharp glass shards from broken windshields. In 1933, cellulose acetate replaced cellulose nitrate as the middle layer in laminated glass, because it was discovered that cellulose nitrate turned yellow after several years' exposure to sunlight. Cellulose acetate was later replaced by polyvinyl butyral (PVB) in 1939. PVB is still used today in the modern manufacturing of laminated safety glass.

FIGURE 13.1

Ford Model A

Recognize, Recall, and Reflect

1.   What was the surprising observation Benedictus made in his lab after dropping a flask?

2.   What had happened to the flask to make it react the way it did?

3.   What was the application Benedictus thought of for his accidental discovery, and how did he come up with this idea?

Investigate

In this activity, you will explore the unique features of three different types of glass: ordinary glass, tempered glass, and laminated glass. First, you will watch the video "Understanding Different Types of Glass" (*www.youtube.com/watch?v=q5dqC0J2UrM*) to observe how the different types of glass break when they come in contact with a destructive force. After seeing the video, your teacher may show you different types of broken glass up close. Then you will answer the questions below.

Safety Note: Do not touch the broken pieces of glass presented in the demonstration.

Questions for Reflection

1. What did you observe about the appearance of the unbroken sheets of ordinary (annealed) glass, tempered safety glass, and laminated safety glass?

2. What are the similarities and differences between the ways in which the three types of glass behave when hit with an impacting object?

3. Which type of glass breaks into small pieces that are less likely to cut you?

4. Which type of glass stays in a sheet when it breaks?

5. Describe an experiment you would like to carry out to learn more about one or more of the types of glass. What question are you asking, and how will your experiment provide an answer to your question?

Apply and Analyze

Safety glass is a general term used to describe glass that has been manufactured to have special properties that reduce the chances of it breaking. Two types of safety glass are laminated glass and tempered glass.

Laminated safety glass is made from two sheets of ordinary glass with a very thin sheet of clear plastic (PVB) in between. A sheet of laminated safety glass resists breaking apart because the middle PVB layer helps to soften the blow of an impacting object while also holding the sharp pieces of broken glass in place. This makes it much less likely that someone will be cut by large shards of broken glass and also makes it more difficult for something (e.g., an impacting object, a passenger in a car) to pass through the sheet of glass. (See Figure 13.2.)

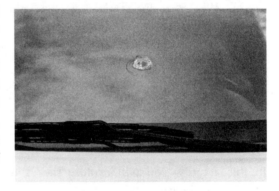

FIGURE 13.2

Broken Laminated Safety Glass

Tempered safety glass is produced when a sheet of glass is heated to near its melting point (about 700°C) and then quickly cooled by blowing cold air onto its surfaces. When glass is cooled in this way, the outside layer hardens first, which determines the size of the final glass sheet. As the middle layers of the glass cool next, they pull on the outer layers. This process creates a stronger glass sheet: To break it would require a force that's generally about four times stronger than the force needed to shatter an ordinary glass sheet of the same thickness. Due to its internal structure, when a tempered glass sheet breaks, the entire sheet breaks into small pieces of glass that are less likely to injure someone. (See Figure 13.3.)

FIGURE 13.3

Broken Tempered Safety Glass

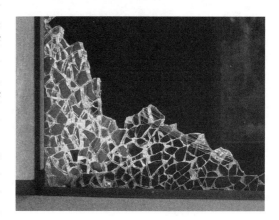

Laminated glass and tempered glass have different advantages and disadvantages. And these advantages and disadvantages help determine how each one is used. For example, laminated glass is easier to crack than tempered glass. But it is harder to puncture or break through a sheet of laminated glass. Therefore, laminated glass is often used when additional protection is needed to prevent something from falling through glass. It can be found in windows, skylights, and balconies. A laminated glass with a thicker middle layer of PVB is called impact glass. This product is often used for windows that need to withstand hurricane-force winds and flying debris. An extra-thick laminated glass made from multiple layers of glass and PVB is called bullet-resistant glass. This is used in bank teller windows and in windshields for aircraft, tanks, and other specialized vehicles. The PVB layer also has the added benefit of reducing the ultraviolet light that passes through the glass.

While laminated glass is commonly used for the front windshield of cars, the side and back windows of cars are often made from tempered glass. Tempered glass is also commonly used for glass components in refrigerators and ovens, fireplace screens, shower and bathtub enclosures, patio doors, and doors and windows in commercial buildings.

Based on what you've learned, answer the following questions:

1. Name a use for tempered glass and a use for laminated glass. Why is that type of glass a good choice for the application?

2. Are there any applications you would *not* choose for tempered glass? For laminated glass? Why not?

3. Why do you think the front windshield in cars is made from laminated glass but the side and rear windows are made from tempered glass?

4. The ability to resist stronger impacts, like the impact from a fired bullet, comes from increasing the number of layers in laminated safety glass.

Should all laminated safety glass products be made to resist the impact from a bullet? Discuss the trade-off between cost and benefit when making your recommendation.

Design Challenge

Engineering is the application of scientific understanding through creativity, imagination, problem solving, and the designing and building of new materials to address and solve problems in the real world. You will be asked to take the science you have learned in this case and design a process or product to address a real-world issue of your choosing.

Engineers use the engineering design process as steps to address a real-world problem (see Figure 13.4). You will now use this process as you come up with a new application for safety glass. In this case, you are asking the question (Step 1) of how safety glass could be used for new purposes. Drawing on your creativity, you will then brainstorm (Step 2) a new way for safety glass to be used in order to solve a problem. Afterward, you will create a plan (Step 3) for this new product. Although you will not actually produce your product, you will create a sketch and/or a model of it (Step 4). Then, you will work with your classmates to think about how you would test (Step 5) and refine (Step 6) your product.

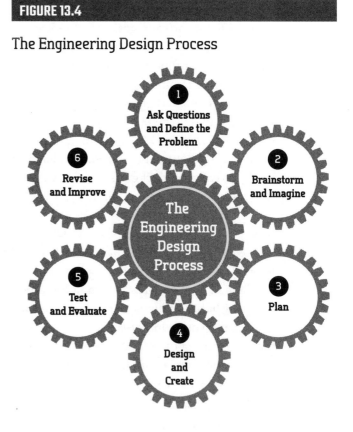

FIGURE 13.4

The Engineering Design Process

1. Ask Questions

Based on your research, consider what new problems may be addressed or what new products could be created using a type of glass that has specific safety features. What problem do you think could be solved with safety glass?

2. Brainstorm and Imagine

Consider a new application for safety glass. You may choose to design a product made from laminated glass, tempered glass, or both. Think about applications for a material that is strong while also being transparent, or where you need a type of glass that is less likely to harm people when it breaks. Also think of how other features of each glass type might be useful. (For example, some people have skin that is sensitive to ultraviolet radiation and must limit their exposure to sunlight. Perhaps a product can be designed from laminated glass to help shield people from the Sun's powerful rays while keeping them safe from collisions or falling objects.) What useful product would you design using safety glass?

3. Create a Plan

Create a plan for a using safety glass technology. Consider: (1) What is the purpose of the product? (2) What are benefits of the product? (3) What are the limitations of the product? Use the Product Planning Graphic Organizer (p. 207) to help you.

4. Design and Create

Consider the following questions and considerations for your safety glass product and its design.

- Which safety glass will your product use?

- How would incorporating safety glass make this product better?

- Would there be limitations or drawbacks to using safety glass, and how would you overcome them?

Now, create a sketch of your product. Make sure your design incorporates the research and exploration you've done.

5. Test and Evaluate

Working with your classmates, come up with a way to test your design to see its effectiveness.

6. Revise and Improve

Give your plans to one of your classmates for review. Listen to his or her feedback on your design. What are some ways you can use the input to refine your design? Take some time to revise and make improvements.

Reflect

1. What technologies might need to be developed to create or manufacture this design?

2. What are any constraints or drawbacks you can foresee with implementing this design?

3. Would there be any environmental or human health concerns to using this product?

Product Planning Graphic Organizer

Proposed Product Idea	
Pros (Benefits)	**Cons (Limitations)**

SAVING LIVES THROUGH AN ACCIDENT

SAFETY GLASS

A Case Study Using the Discovery Engineering Process

Lesson Overview

In this lesson, students will learn about ordinary glass and two types of glass (laminated and tempered) that have unique safety characteristics. After reading about the accidental discovery of safety glass in a chemistry laboratory, students will observe what happens when the different types of glass break. (There is an option to carry out this demonstration in the classroom or to watch a video of the breaking glass.) Students will compare and contrast the breaking patterns and will brainstorm applications for different glass products.

Lesson Objectives

By the end of this case study, students will be able to

- Describe the characteristics of two types of safety glass (laminated and tempered).

- Analyze the benefits and limitations of two types of safety glass as a material for consumer products.

- Design a new application for safety glass that solves a problem.

The Case Study Approach

This lesson uses a case study approach. Explaining the purpose of case studies will encourage your students to relate to the material and engage with the problem. At the heart of each case study in this book is a true story, one that describes how someone in his or her everyday life or during a routine workday made an observation or did a simple experiment that led to a new insight or discovery. Case studies are designed to get students actively engaged in the process of problem solving. The narrative of the case supplies authentic details that place the student in the role of the inventor and provide scaffolds for critical thinking and deep reflection. A case is more than a paragraph to read or a story to analyze but rather a way of framing problems, synthesizing what is known, and thinking creatively about new applications and solutions. In this lesson, students consider how safety glass was discovered and work together to think about new applications for this glass to solve real-life problems.

Use of the Case

Due to the nature of these case studies, teachers may elect to use any section of each case for their instructional needs. The sections are sequenced in order (scaffolded) so students think more deeply about the science involved in the case and develop an understanding of engineering in the context of science.

Curriculum Connections

Lesson Integration

You could use this case as a way to integrate engineering into a lesson on the sub-atomic structure of crystalline and amorphous solids, properties of glass, or even projectiles and collisions.

Related Next Generation Science Standards

PERFORMANCE EXPECTATIONS

- MS-PS1-3. Gather and make sense of information to describe that synthetic materials come from natural resources and impact society.

- HS-ETS1-2. Design a solution to a complex real-world problem by breaking it down into smaller, more manageable problems that can be solved through engineering.

- HS-ETS1-3. Evaluate a solution to a complex real-world problem based on prioritized criteria and trade-offs that account for a range of constraints, including cost, safety, reliability, and aesthetics, as well as possible social, cultural, and environmental impacts.

SCIENCE AND ENGINEERING PRACTICES

- Asking Questions and Defining Problems

- Analyzing and Interpreting Data

- Engaging in Argument From Evidence

- Constructing Explanations and Designing Solutions

CROSSCUTTING CONCEPTS

- Cause and Effect
- Structure and Function

Related National Academy of Engineering Grand Challenges

- Restore and Improve Urban Infrastructure
- Engineer the Tools of Scientific Discovery

Lesson Preparation

You will need to make copies of the entire student section for the class. Look at the Teaching Organizer (Table 13.1) for suggestions on how to organize the lesson.

During the Investigate section, you will need internet access in order to show the 10-minute-long video "Understanding Different Kinds of Glass" (*www.youtube.com/watch?v=q5dqC0J2UrM*). This video demonstrates how five different types of glass (100-year-old antique glass, annealed glass, tempered glass, laminated glass, and impact glass [a thicker type of laminated glass]) react when hit with an impacting object. If time is a factor, you can stop the video about six and a half minutes in, after the discussion of laminated glass.

If you wish to provide examples of broken glass for students to observe, prepare the glass ahead of time without students present. Glass sheets may be purchased from most hardware stores. Local building companies or contractors may also have extra pieces that would be suitable for this demonstration. Start by clearing away unnecessary objects and laying down a tarp or heavy-duty trash bags to catch any projectile pieces of glass. Set up the tarp or bags so that a section is hanging vertically to prevent glass from scattering backward from impact. Lay down the sheets of glass. Cover the glass with a large, heavy-duty plastic bag. Use the hammer to break the glass sheets. Lay out the broken pieces of glass for students to see. Properly dispose of the broken glass once the demo is complete. Note that you should wear contractor gloves and safety goggles at all times during the preparation, presentation, and cleanup of the demonstration.

Materials

For teacher demonstration:

- 1 sheet of ordinary glass
- 1 sheet of tempered glass

- 1 sheet of laminated glass

- 1 hammer

- Heavy-duty/contractor-strength trash bags

- Tarp or additional trash bags

- Contractor gloves

- Safety goggles

Safety Note for Students: Do not touch the broken pieces of glass presented in the demonstration.

Time Needed

70 minutes

TABLE 13.1

Teaching Organizer

Section	Time Suggested	Materials Needed	Additional Considerations
The Case	15 minutes	Student packet	Could be read in class or as a homework assignment prior to class
Investigate	10 minutes	Student packet; internet access; for teacher demonstration: 1 sheet of ordinary glass, 1 sheet of tempered glass, 1 sheet of laminated glass, 1 hammer, heavy-duty/contractor-strength trash bags, tarp or additional trash bags, contractor gloves, safety goggles	Whole-class activity
Apply and Analyze	15 minutes	Student packet	Small-group or individual activity
Design Challenge	30 minutes	Student packet	Small-group activity

Teacher Background Information

Read the information below for details and background information about the various forms of glass that can be found.

Glass: Glass is a solid material that has an amorphous structure at the atomic level, which means it contains little or no areas of ordered crystalline structure. Glass sheets often undergo different types of processing to make them stronger before they are used in products. For more information about glass to share with your class, watch the TED-Ed video "Why is glass transparent," which explains general properties of glass. You can find the video here: *https://ed.ted.com/lessons/why-is-glass-transparent-mark-miodownik*.

Annealed Glass: Annealing occurs when a glass sheet is heated to near its melting point and slowly cooled. Many materials form more ordered structures at the atomic level when they are heated and then slowly cooled. In glass, this process is applied to help reorder sites in the glass that may be weaker and highly likely to fracture due to their extremely disordered structure. Annealing doesn't reorder all the atoms into a crystalline structure; rather, the process may be thought of as repairing major sites of dissymmetry. Annealed glass sheets break into large, sharp shards, which may be dangerous.

Tempered Glass: Tempered safety glass is produced when a sheet of glass is heated to near its melting point (about 700°C) and then quickly cooled by blowing cold air onto its surfaces. When glass is cooled in this way, the outside layer hardens first, which determines the size of the final glass sheet. As the middle layers of the glass cool next, they pull on the outer layer of the glass. This creates a stronger glass sheet: To break it would require a force that's generally about four times stronger than the force needed to shatter an ordinary glass sheet of the same thickness. Due to its internal structure, when a tempered glass sheet breaks, the entire sheet breaks into small pieces of glass that are less likely to injure someone.

Laminated Glass: Laminated safety glass is made from two sheets of glass (often annealed glass) with a very thin sheet of clear plastic (PVB) in between. A sheet of laminated safety glass resists breaking apart because the middle PVB layer helps to soften the blow of an impacting object while also holding the sharp pieces of broken glass in place. This makes it much less likely that someone will be cut by large shards of broken glass and also makes it more difficult for something (e.g., an impacting object, a passenger in a car) to pass through the sheet of glass. A sheet of laminated glass cracks more easily than a sheet of tempered glass of the same thickness, but the laminated glass sheet is more likely to hold its shape.

Vocabulary

- annealed
- cellulose nitrate
- laminated
- polyvinyl butyral, PVB
- tempered

Teacher Answer Key

Recognize, Recall, and Reflect

1. **What was the surprising observation Benedictus made in his lab after dropping a flask?**

    Benedictus observed that the flask did not break apart when it hit the floor, despite being dropped from a height of about 11 feet. The flask cracked but was able to maintain its shape, and the broken pieces of glass held together.

2. **What had happened to the flask to make it react the way it did?**

    The flask had held a solution of cellulose nitrate which had dried clear on the inside of the flask. When the glass broke, the cellulose nitrate layer held the glass shards in place.

3. **What was the application Benedictus thought of for his accidental discovery, and how did he come up with this idea?**

    Benedictus didn't think of an application for his new discovery right away. After hearing about multiple automobile accidents where passengers were injured by sharp glass shards, he realized that his new discovery might be a useful material for the manufacture of windshields. He then created a process to sandwich a layer of cellulose nitrate between two sheets of glass and created the product known as laminated safety glass.

Questions for Reflection

1. **What did you observe about the appearance of the unbroken sheets of ordinary (annealed) glass, tempered safety glass, and laminated safety glass?**

    Answers will vary.

2. **What are the similarities and differences between the ways in which the three types of glass behave when hit with an impacting object?**

    Annealed glass will shatter into sharp, irregularly shaped pieces that range in size from small to large. Unlike the two types of safety glass (tempered and laminated), annealed glass can easily cut you if it breaks. Tempered glass, which is harder to break than annealed glass, will shatter into many small pieces. Unlike annealed and tempered glass, laminated glass will not quickly break into many pieces when struck. It takes repeated strikes to break through a piece of laminated glass. Laminated glass will break into irregularly shaped pieces that range in size, similar to annealed glass. But the broken pieces of laminated glass are held together by a thin plastic layer.

3. **Which type of glass breaks into small pieces that are less likely to cut you?**

Tempered safety glass

4. **Which type of glass stays in a sheet when it breaks?**

Laminated safety glass

5. **Describe an experiment you would like to carry out to learn more about one or more of the types of glass. What question are you asking, and how will your experiment provide an answer to your question?**

Answers will vary.

Apply and Analyze

1. **Name a use for tempered glass and a use for laminated glass. Why is that type of glass a good choice for the application?**

Student answers will vary. An example of an application for tempered glass is as a shower door. If broken, tempered glass breaks apart into many small pieces that are less likely to cause injury. So, if someone slips and falls through the glass door of a shower, he or she will be less likely to be cut by broken glass. One use for laminated glass is as external windows in homes located in hurricane zones. The PVB layer in this type of glass helps hold sheets together, even under great force.

2. **Are there any applications you would *not* choose for tempered glass? For laminated glass? Why not?**

Because tempered glass shatters into many small pieces when it breaks, it would not work well for a product like a skylight. Laminated glass would not work well in products that are meant to be broken easily in the event of an emergency (e.g., to access emergency safety equipment like a fire hose).

3. **Why do you think the front windshield in cars is made from laminated glass, but the side and rear windows are made from tempered glass?**

Using laminated glass for the windshield makes it more likely to stay in one piece if hit by a rock or other debris while the car is in motion. And if the windshield stays intact, it's less likely that passengers will be injured by broken glass. The strength of the windshield also supports the car roof and protects passengers from the roof caving in by remaining in place in the event of the car rolling over.

Tempered glass is often less expensive to manufacture than laminated glass and may be used in applications that require safety glass, but where the specific features of laminated glass are not required. If a car passenger needs to exit a vehicle in an emergency, it is easier to break tempered glass than laminated glass.

4. **The ability to resist stronger impacts, like the impact from a fired bullet, comes from increasing the number of layers in laminated safety glass. Should all laminated safety glass products be made to resist the impact from a bullet? Discuss the trade-off between cost and benefit when making your recommendation.**

   Answers will vary depending on what each student recommends. Students should address the costs and benefits of using stronger glass in their recommendations. For instance, a student might acknowledge that using extra layers to create bulletproof glass would make windows more expensive. However, the student might argue that the benefit of safety is more important than this cost.

Reflect

1. **What technologies might need to be developed to create or manufacture this design?**

   Student answers may vary depending on their design.

2. **What are any constraints or drawbacks you can foresee with implementing this design?**

   Student answers may vary depending on their design.

3. **Would there be any environmental or human health concerns to using this product?**

   Student answers may vary depending on their design.

Assessment

The Design Challenge can be assessed using the rubric in the appendix (p. 377). You can also ask students to take and share pictures of glass in use at home, at school, or in their communities. Then have students form groups to discuss the photos and guess what types of glass are pictured. Ask students to provide evidence for their reasoning.

Extensions

This lesson can be followed with lessons about the subatomic structure of crystalline and amorphous solids, properties of glass, or projectiles and collisions.

Resources and References

Carraher, C. E. 2013. *Introduction to polymer chemistry.* 3rd ed. Boca Raton, FL: CRC Press.

The Craftsman Blog. 2016. "Understanding different types of glass." YouTube video. *www.youtube.com/watch?v=q5dqC0J2UrM.*

Editors of Encyclopaedia Britannica. 2016. "Nitrocellulose." In *Encyclopaedia Britannica. www.britannica.com/science/nitrocellulose.*

Garrett, A. B. 1963. *A flash of genius.* Princeton, NJ: D. Van Nostrand Company, Inc.

Miodownik, M. 2014. "Why is glass transparent?" TED-Ed video. *https://ed.ted.com/lessons/why-is-glass-transparent-mark-miodownik.*

Neiger, C. 2009. How automotive glass works. *https://auto.howstuffworks.com/car-driving-safety/safety-regulatory-devices/auto-glass.htm.*

A WEARABLE WATER FILTER

The Sari

A Case Study Using the Discovery Engineering Process

Introduction

Cholera affects more than 150,000 people worldwide each year. It is an infectious disease of the intestines that causes severe diarrhea, often leading to dehydration and even death if it is not treated. Cholera was common in the United States in the 1800s, but modern water and sewage treatment has reduced the risk of the disease. While cholera decreased in the United States, it is still problematic in many parts of the world. This is especially true in places where people drink from open, shared water sources. In this lesson, you will learn how one country is attempting to reduce the risk of cholera through a simple tool: the sari (Figure 14.1).

FIGURE 14.1

Women Wearing Saris

Lesson Objectives

By the end of this case study, you will be able to

- Describe the process through which cholera is spread.

- Analyze the effectiveness of water filtration using a sari versus other materials.

- Design a water filtration system.

The Case

Read the following summary of the CNN article "The Link Between Saris and Cholera," published on June 5, 2013. The article features Dr. Rita Colwell, who has been studying cholera for more than five decades. Colwell's research has helped protect many families in Bangladesh from cholera.

Rita Colwell spent more than 50 years studying cholera. In the 1970s, she and her team discovered that cholera is caused by a bacterium naturally found in water. The bacterium attaches itself to the interior of the intestine after it has been ingested. Once it is settled, it begins producing a toxin that leads to vomiting and diarrhea. If an infected person does not receive medical attention and lots of fluids, he or she can become dehydrated, which can lead to death. Treatments of cholera include antibiotics and rehydration.

Cholera bacteria can be killed by boiling water before drinking it or using it to prepare food. However, in some parts of the world, wood or other types of fuel are not readily available to make fires for boiling water. That means other solutions are needed.

During her research, Colwell found that cholera is associated with a type of plankton (or microscopic organism) called a copepod. As women in Bangladesh often filter their drinks at home with a piece of sari cloth to remove leaves and insects, Colwell decided to test whether the sari could be effective for also filtering out the copepods that carry cholera.

The sari—a garment often worn by women in India, Bangladesh, Nepal, and Sri Lanka—is usually around six feet long. In the lab, Colwell found that a sari folded four to eight times was able to filter out all of the zooplankton (plankton composed of animals), and most of the phytoplankton (plant plankton), along with the cholera bacteria that were attached to plankton or other pieces of material floating in the water.

To test whether this was an effective way to prevent cholera, Colwell conducted a three-year study in Bangladesh in the early 2000s. She worked with 50 villages and more than 150,000 people. The scientist and her team went to the villages and taught the women about the importance of filtering water, why filtering helps keep people healthy, and how to effectively filter harmful bacteria from the water. The team then measured the impact of their work and found that the number of people with cholera dropped by 50%. When the research team went back in the years after the study was complete, they found that 75% of the women were still filtering their water.

Recognize, Recall, and Reflect

1. Why does cholera make people sick?

2. What are ways you can remove cholera from water?

3. What happened when Colwell trained women in Bangladesh to filter bacteria from their water with sari fabric?

Investigate

In this activity, you will explore different types of fabrics and their effectiveness in filtering water.

Materials

For each group of students:

- 1 piece of cotton fabric

- 1 piece of wool fabric

- 1 piece of nylon fabric or nylons

- 1 piece of other fabric (silk, bamboo, polyester, etc.)

- 1 hand lens

Safety Note: If you are allergic to wool fabric, do not handle it. Wash your hands with soap and water immediately after completing this activity.

Create, Innovate, and Investigate

- Begin by looking at the magnified photos of fabric (Figures 14.2–14.4, p. 220). How is each one different?

- Examine the different types of actual fabrics. Feel each fabric. Stretch each fabric. Look at each fabric under the hand lens.

- What are similarities and differences you observe when you compare the fabrics? List your observations in the Characteristics of Fabric Chart (p. 221).

- Based on your investigation, what can you conclude about each fabric's usefulness as a water filter? List these thoughts underneath the Characteristics of Fabric Chart.

FIGURE 14.2

Yarn

FIGURE 14.3

Wool

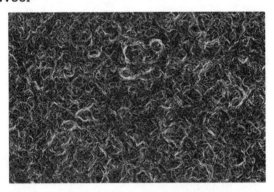

FIGURE 14.4

Nylon

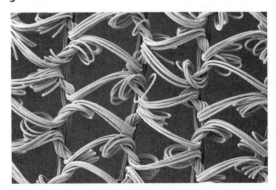

Characteristics of Fabric Chart

Fabric Type	Observations	Comments on Usefulness in Filtering Water
Cotton		
Wool		
Nylon		
Other fabric		

Of the fabrics we examined, we believe _____ would be the best option for filtering water because:

Questions for Reflection

1. What did you observe about the different photos of magnified fabrics?

2. What did you observe about the different types of actual fabrics?

3. How did each actual fabric look under the hand lens?

4. Which of these fabrics do you think would be easiest to find in developing countries?

5. Based on your examination, which fabric would be the best to use as a water filter?

Apply and Analyze

When testing their fabric filters, Dr. Colwell's team investigated two different types of fabric to find out which one was more effective. They trained one group of towns to filter their water using sari fabric folded four times (for a total of 16 layers). Another group of towns was taught to filter their water with nylon fabric. A third group of towns acted as the control and were not taught how to filter their water with fabric. There were 27 villages that used the sari method, 25 villages used nylon, and 13 villages were in the control group. The researchers gathered data about the villages' rates of cholera in 1997 and 1998, and then they began the study in 1999. Look at Figure 14.5 to find out what the researchers observed.

FIGURE 14.5

Cholera Cases by Year

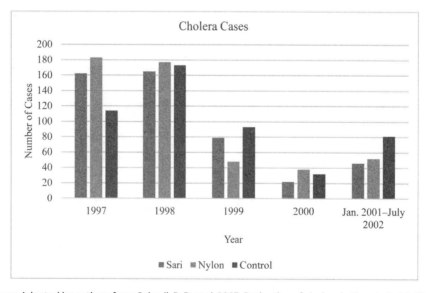

Source: Adapted by authors from Colwell, R. R. et al. 2003. Reduction of cholera in Bangladeshi villages by simple filtration. *Proceedings of the National Academy of Sciences* 100 (3): 1051–1055.

1. Which group saw the greatest reduction in the number of cholera cases?

2. What are some reasons the number of cholera cases could have increased after the year 2000?

3. Which material would you recommend villages use for filtering? Why?

Design Challenge

Engineering is the application of scientific understanding through creativity, imagination, problem solving, and the designing and building of new materials to address and solve problems in the real world. You will be asked to take the science you have learned in this case and design a process or product to address a real-world issue of your choosing.

Engineers use the engineering design process as steps to address a real-world problem (see Figure 14.6). You will now use this process as you come up with a new filtration method. In this case, you are asking the question (Step 1) of how you can design a new filtration method. Drawing on your creativity, you will then brainstorm (Step 2) a new product that filters water for

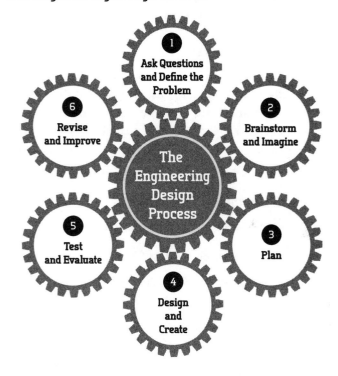

FIGURE 14.6

The Engineering Design Process

1. Ask Questions and Define the Problem
2. Brainstorm and Imagine
3. Plan
4. Design and Create
5. Test and Evaluate
6. Revise and Improve

The Engineering Design Process

drinking. Afterward, you will create a plan (Step 3) for this new product. Next, you will create a sketch of your product design, and then build it (Step 4). Then, you will work with your classmates to think about how you would test (Step 5) and refine (Step 6) your product.

1. Ask Questions

Based on your previous research, consider a reason why you might need to filter water. (One example is if you ran out of water while hiking and needed to drink from a stream). Identify what you are trying to remove from the water. How small are the objects?

2. Brainstorm and Imagine

After thinking of a situation in which you would need to filter water, brainstorm a type of water filter that you could use in this scenario. Think about the materials needed to build this filter. Get creative! In addition to fabric, what materials have pores, or small openings, that would allow water through but would filter out impurities? Make a list of the items you need. (Note that a traditional water filter is built by layering gravel on top of sand over activated carbon, which is used to remove chemicals. For more ideas, explore the Center for Disease Control's website on filtration systems: *www.cdc.gov/healthywater/drinking/home-water-treatment/household_water_treatment.html*.)

3. Create a Plan

Create a plan for building your water filter. Double check that you have listed all the necessary materials. How should the materials be layered or positioned? What structural items will you need to hold your filter together? Also make sure you design a collection device to gather the water once it has passed through your filter. In devising your plan, think of the pros and cons of your design. Fill out the Product Planning Graphic Organizer (p. 226) to help you make your plan.

4. Design and Create

Consider the following questions and considerations for your water filter.

- How would this filter work?

- Are there any limitations or drawbacks to your product? If so, how would you overcome them?

- What are any constraints or drawbacks you can foresee with implementing this design?

- Would there be any safety concerns regarding your water filter–based product?

After considering these questions, create a sketch of your filter that includes all of its parts. Create a diagram illustrating the direction water will move into and through the filter, and show what will be removed in each stage. Once your sketch is complete, collect the materials needed to construct your water filter. (Your teacher will let you know whether materials will be provided in class or whether you will need to bring them in from home.) Then build your product.

5. Test and Evaluate

After you have built your water filter, you will get a water sample from your teacher. Carefully pour the water through your filter. Once it has finished filtering, compare your filtered sample to an unfiltered sample of water from the same source, using visual inspection or by comparing the two samples with a hand lens or microscope. How effective was your filter?

Safety Note: Do not drink your water sample even after it has been filtered.

6. Revise and Improve

After comparing your water to the original sample, consider how you might improve your filter. Are there items you could add or remove to make it more effective? Consider other ways to improve it such as making it more portable or faster.

Reflect

1. What worked well in your design?

2. What are the limitations or drawbacks of your design?

3. What environmental issues might you have to take into consideration when designing a water filter?

Product Planning Graphic Organizer

Proposed Product Idea

Pros (Benefits)	Cons (Limitations)

A WEARABLE WATER FILTER

THE SARI

A Case Study Using the Discovery Engineering Process

Lesson Overview

In this lesson, students explore how researchers developed a new solution to cholera outbreaks in some regions of the world. After watching women in Bangladesh use their saris to filter insects and leaves out of their drinking water, researchers discovered the same process could be used to remove cholera from the water. Students will learn how cholera affects people and how researchers are attempting to prevent cholera infections. They will test different types of fabric to use as filters and examine data from the sari study. Then, they will build their own water filter.

Lesson Objectives

By the end of this case study, students will be able to

- Describe the process through which cholera is spread.

- Analyze the effectiveness of water filtration using a sari versus other materials.

- Design a water filtration system.

The Case Study Approach

This lesson uses a case study approach. Explaining the purpose of case studies will encourage your students to relate to the material and engage with the problem. At the heart of each case study in this book is a true story, one that describes how someone in his or her everyday life or during a routine workday made an observation or did a simple experiment that led to a new insight or discovery. Case studies are designed to get students actively engaged in the process of problem solving. The narrative of the case supplies authentic details that place the student in the role of the inventor and provide scaffolds for critical thinking and deep reflection. A case is more than a paragraph to read or a story to analyze but rather a way of framing problems, synthesizing what is known, and thinking creatively about new applications and solutions. In this lesson, students consider how fabric can be used as a water filter and come up with their own water filtration system that can be used to solve real-life problems.

Use of the Case

Due to the nature of these case studies, teachers may elect to use any section of each case for their instructional needs. The sections are sequenced in order (scaffolded) so students think more deeply about the science involved in the case and develop an understanding of engineering in the context of science.

Curriculum Connections

Lesson Integration

You could use this case as a way to integrate engineering into a lesson on water as the "universal solvent" or chemical versus physical changes. It may be useful to review the properties of water with these resources from the National Science Teachers Association:

- Book chapter on water's dissolving ability
  https://common.nsta.org/resource/?id=10.2505/9781936959020.5a

- Journal article on the Dirty Water Challenge
  https://common.nsta.org/resource/?id=10.2505/4/sc07_044_09_26

- Journal article on how human actions affect water quality and quantity
  https://common.nsta.org/resource/?id=10.2505/4/ss08_031_06_26

Related Next Generation Science Standards
PERFORMANCE EXPECTATIONS

- MS-ETS1-1. Define the criteria and constraints of a design problem with sufficient precision to ensure a successful solution, taking into account relevant scientific principles and potential impacts on people and the natural environment that may limit possible solutions.

- MS-ETS1-2. Evaluate competing design solutions using a systematic process to determine how well they meet the criteria and constraints of the problem.

- HS-ETS1-3. Evaluate a solution to a complex real-world problem based on prioritized criteria and trade-offs that account for a range of constraints, including cost, safety, reliability, and aesthetics, as well as possible social, cultural, and environmental impacts.

SCIENCE AND ENGINEERING PRACTICES

- Analyzing and Interpreting Data

- Engaging in Argument From Evidence

- Constructing Explanations and Designing Solutions

Related National Academy of Engineering Grand Challenge

- Provide Access to Clean Water

Lesson Preparation

You will need to make copies of the entire student section for the class. You will also need to collect fabric samples or ask students to check labels on materials from home and bring in the appropriate fabrics. For the Design Challenge, you can supply a variety of materials for the class (see suggestions in the material list below) or ask the students to supply their own based on their designs. Look at the Teaching Organizer (Table 14.1, p. 230) for suggestions on how to organize the lesson.

Materials

For each group of students:

- 1 piece of cotton fabric

- 1 piece of wool fabric

- 1 piece of nylon fabric or nylons

- 1 piece of other fabric (silk, bamboo, polyester, etc.)

- 1 hand lens

- Items for building water filters (e.g., funnels, gravel, rice, coffee filters, fabric, sand, soda bottles, activated charcoal) may be supplied by students

Safety Note for Students: If you are allergic to wool fabric, do not handle it. Wash your hands with soap and water immediately after completing this activity.

Time Needed

2 class periods

TABLE 14.1

Teaching Organizer

Section	Time Suggested	Materials Needed	Additional Considerations
The Case	5 minutes	Student packet	Could be read in class or as a homework assignment prior to class
Investigate	10 minutes	Student packet, 1 piece of cotton fabric, 1 piece of wool fabric, 1 piece of nylon fabric or nylons, 1 piece of other fabric (silk, bamboo, polyester), 1 hand lens	Small-group activity
Apply and Analyze	10 minutes	Student packet	Small-group or individual activity
Design Challenge	1 class period	Student packet, materials for building water filters (e.g., funnels, gravel, rice, coffee filters, fabric, sand, soda bottles, activated charcoal) (*Note:* You can provide materials for students or ask them to bring in materials from home.)	Small-group activity

Teacher Background Information

Cholera is a waterborne bacterium. It is often carried by copepods in the water. It is frequently a problem in the parts of the world where people share open water sources. According to some researchers, the prevalence of cholera is increasing with climate change and increased populations. Warmer water harbors more cholera and, as populations increase, it is easier for the bacteria to spread. Cholera can be filtered out of water or the water can be treated chemically or through boiling.

Vocabulary

- bacteria
- cholera
- copepod
- phytoplankton
- plankton
- sari
- zooplankton

Teacher Answer Key

Recognize, Recall, and Reflect

1. **Why does cholera make people sick?**

   The bacteria attach to the lining of the intestine and releases toxins. Most people with cholera die due to dehydration.

2. **What are ways you can remove cholera from water?**

   Boiling, chemical treatment, or filtration

3. **What happened when Dr. Colwell trained women in Bangladesh to filter bacteria from their water with sari fabric?**

   The number of cholera cases decreased.

Questions for Reflection

1. **What did you observe about the different photos of magnified fabrics?**

   Answers will vary.

2. **What did you observe about the different types of actual fabrics?**

   Answers will vary.

3. **How did each actual fabric look under the hand lens?**

   Answers may vary but could, for example, discuss how cotton weave is tighter than nylon. Answers could also include how the shapes of each fabric's fibers are different.

4. **Which of these fabrics do you think would be easiest to find in developing countries?**

   The fabrics made from natural fibers, like cotton and wool

5. **Based on your examination, which fabric would be the best to use as a water filter?**

   Answers will vary.

Apply and Analyze

1. **Which group saw the greatest reduction in the number of cholera cases?**

   The group using nylon, followed by the group using saris

2. **What are some reasons the number of cholera cases could have increased after the year 2000?**

   Answers may vary but could include a decrease in the number of people filtering, temperature changes, more bacteria in the water, etc.

3. **Which material would you recommend villages use for filtering? Why?**

   Answers will vary. Students may say nylon, because it filtered out the most bacteria. Or they may recommend the sari, because it was still effective and more readily available.

Reflect

1. **What worked well in your design?**

   Answers will vary.

2. **What are the limitations or drawbacks of your design?**

   Answers will vary.

3. **What environmental issues might you have to take into consideration when designing a water filter?**

   Answers will vary.

Assessment

You can give students a completion grade for creating and testing a water filter. However, you could also have students write up a lab report describing their design and giving explanations for why their filter did or did not work.

Extensions

This lesson can be followed with an activity from Project WET called Poison Pump (*www.rivanna-stormwater.org/bacteria.pdf*). It discusses the outbreak of cholera in London and how scientists used data visualization to identify the source of the infection.

Resources and References

Aretxabaleta, A., G. Brooks, and N. West. 2011. *Project earth science: Physical oceanography.* 2nd ed. Arlington, VA: NSTA Press.

CDC. 2014. A guide to drinking water treatment technologies for household use. *www.cdc.gov/healthywater/drinking/home-water-treatment/household_water_treatment.html.*

Colwell, R. R., A. Huq, M. S. Islam, K. M. A. Aziz, et al. 2003. Reduction of cholera in Bangladeshi villages by simple filtration. *Proceedings of the National Academy of Sciences* 100 (3): 1051–1055.

Gordon, J. 2008. How do our actions affect water quantity and quality? *Science Scope* 31 (6): 26–31.

Huq, A., M. Yunus, S. S. Sohel, and A. Bhuiya. 2010. Simple sari cloth filtration of water is sustainable and continues to protect villagers from cholera in Matlab, Bangladesh. *MBio* 1 (1): e00034–10.

Murray, K. 2013. The link between saris and cholera. *www.cnn.com/2013/06/05/health/lifeswork-colwell-cholera/index.html*.

Schluter K., M. Walker, and A. Kremer. 2007. The Dirty Water Challenge. *Science and Children* 44 (9): 26–29.

WebMD. Facts about cholera. *www.webmd.com/a-to-z-guides/cholera-faq#1* (accessed January 10, 2018).

FROM SHIP TO STAIRCASE

The History of the Slinky

A Case Study Using the Discovery
Engineering Process

Introduction

The Slinky is a popular toy that can perform many fun actions like "walking" down a flight of stairs or appearing to hover in mid-air for a short time when dropped. Since 1945, millions of Slinkys have been sold around the world, making it an iconic, successful toy that many children and adults enjoy. A Slinky is a coil spring (also called a helical spring) that looks somewhat similar to a strand of curly hair when extended. It's made of metal or plastic and comes in many different colors. Coil springs absorb and store energy. They can be found in a variety of products, including mattresses and toasters. The Slinky has been used by everyone from science teachers to NASA officials to explain science and engineering topics, such as energy and gravity.

Lesson Objectives

By the end of the case study, you will be able to

- Describe the characteristics of coil springs.

- Analyze the benefits and limitations of coil springs as a device to store energy.

- Design a new application for coil springs that solves a current problem.

The Case

Read the following summary about the history of the Slinky. This account outlines the accidental discovery of the Slinky and how it was transformed into one of the most popular children's toys in recent history. Once you are finished reading, answer the questions that follow.

Richard James was a mechanical engineer who worked for the U.S. Navy in the 1940s. During World War II, his job was to develop ideas for protecting sensitive instruments on large ships stuck in turbulent, rough storms. James was focused on attaching sensitive instruments (for instance, instruments that track time or a ship's location) to the ship's interior with coiled springs, which absorb kinetic energy from the moving vessel and keep the instruments safe and in place. He wanted to determine which coil spring worked the best.

While experimenting with different coils in 1943, James accidentally knocked one of them off a table. The coil spring bounced and "walked" around the room. Fascinated by the idea that a coil spring could walk, James experimented with different types of metals until he created a toy coil spring. James's wife, Betty James, was also excited by the new toy, and she is credited with naming it the Slinky.

Richard and Betty James requested a loan from a bank to help make a set of Slinky toys. Toy sales were low and some toy stores did not want to feature the Slinky. In 1945, however, one store allowed the couple to show demonstrations of the Slinky in action. The demonstration was a hit, and the store sold out of its entire stock of Slinkys by the end of the day! Because of its success, other types of Slinky toys were developed, including the famous Slinky Dog.

Recognize, Recall, and Reflect

1. What was Richard James's job during World War II?

2. Why are coil springs effective at keeping sensitive instruments in ships safe?

3. Why do you believe the Slinky was not a toy that most stores initially wanted to sell? Why do you think the demonstration of the Slinky helped it become a favorite children's toy?

Investigate

In this activity, you will observe three demonstrations given by your teacher of a toy coil spring. Record your observations in the Coil Spring Demonstration Chart below. In recording your observations, use these questions as a guide:

- What did the coil spring do?
- What did you notice?
- What do you wonder?
- What was surprising?
- What questions do you have about the movement of the coil spring?

Coil Spring Demonstration Chart

Demonstration Description	Observations	Questions
Demonstration #1: Toy coil spring "walking" down stairs or an inclined plane		
Demonstration #2: Toy coil spring dropped in mid-air several feet from the ground		
Demonstration #3: Toy coil spring creating waves		

Once the demonstration is complete, you will explore the movement and properties of a toy coil spring with the activities below. After gathering the needed materials, read each activity description in the Create, Innovate, and Investigate section. Then document your observations or responses to questions on your Student Observation and Investigation Chart (pp. 240–241).

Materials

For each group of students:

- Toy coil spring

- An inclined plane or stairs

- Paperclips (1 per student)

- Pencils or other cylindrical objects (1 per student)

- Safety glasses or goggles (1 pair per student)

Safety Note: Use caution not to trip over the coil spring on the floor. Wash your hands with soap and water immediately after completing this activity.

Create, Innovate, and Investigate

- **Activity #1:** Begin by examining the toy coil spring. What do you observe (see, feel, and/or hear)? Record your observations.

- **Activity #2:** How is this coil spring similar to and different from other springs you have seen? Record your answers.

- **Activity #3:** Place your coil spring on an inclined plane or stairs and observe the movement. What do you notice? Record your observations.

- **Activity #4:** Drop your coil spring in mid-air and observe the movement. What do you notice? Record your observations.

- **Activity #5:** Create a wave with your coil spring. What do you notice? Record your observations.

- **Activity #6:** Think of a different way for the coil spring to move. Record detailed instructions to allow other students to replicate, or copy, your new coil spring movement. What do you notice about the new movement of the product? Record your detailed instructions and observations.

- **Activity #7:** Try out one of your classmate's detailed instructions for a new coil spring movement. Record your observations.

- **Activity #8:** Explore and investigate various objects that use springs to work. What is the purpose of their springs? Record your findings.

- **Activity #9:** Design your own coil spring using a paperclip. (Straighten the paperclip and wind it around a pencil or another cylindrical object. Then slide the now-coiled spring off the object.) Make observations as you play with the spring. Record your findings.

Student Observation and Investigation Chart

Activity Description	Guiding Questions	Questions About the Activity
Activity #1: **Examine the coil spring.**	What do you SEE, FEEL, and HEAR? Record your observations below.	Write down any questions you have.
Activity #2: **Compare the coil spring to others.**	How is the coil spring similar to and different from other springs you have seen? Record your observations below.	Write down any questions you have.
Activity #3: **Make your coil spring "walk" down the stairs or an inclined plane.**	What do you notice about the movement of the coil spring? Record your observations below.	Write down any questions you have.
Activity #4: **Drop the coil spring in mid-air from several feet above the ground.**	What do you notice about the movement of the coil spring? Record your observations below.	Write down any questions you have.
Activity #5: **Create waves with the coil spring.**	What do you notice about the movement of the coil spring? Record your observations below.	Write down any questions you have.

Continued

Student Observation and Investigation Chart (*continued*)

Activity Description	Guiding Questions	Questions About the Activity
Activity #6: **Come up with a different way to move the coil spring. Write down instructions to show your classmates how to create this movement.**	Record instructions on how to make your new coil spring movement below. What do you notice about this new movement of the coil spring? Record your observations below.	Write down any questions you have.
Activity #7: **Try out a classmate's instructions to move the coil spring in a new way.**	What do you notice about this new movement of the coil spring? Record your observations below.	Write down any questions you have.
Activity #8: **Explore and investigate various objects that use springs to work. List the objects and the purpose of their springs.**	List the objects you found along with the purpose of their springs below.	Write down any questions you have.
Activity #9: **Design your own coil spring using a paperclip.**	What do you SEE, FEEL, and HEAR? Record your observations below.	Write down any questions you have.

Questions for Reflection

1. What did you observe about the movement and properties of a coil spring?

2. What is unique about the coil spring?

Apply and Analyze

In this case study, you learned that coil springs absorb energy to protect items or store energy so that an item can use the energy at a later time. Coil springs are used for many everyday applications like cars and pens. They're also used in cutting-edge research. For instance, one area of research focuses on how medical devices designed as coiled structures help treat illnesses. Go to the following link and read the story about coil implants used to treat lung disease: *www.foxnews. com/health/2016/01/12/implanted-coils-help-some-lung-disease-patients-study-says.html*. Then answer the questions below.

1. What is chronic obstructive pulmonary disease (COPD)?

2. How do coils help treat individuals who have COPD?

3. What are the benefits of using coils to treat COPD?

4. According to the article, what were the challenges or downsides of using coils during the study? Why should researchers test the coils on a study group before allowing all doctors to use coils in their patients?

Design Challenge

Engineering is the application of scientific understanding through creativity, imagination, problem solving, and the designing and building of new materials to address and solve problems in the real world. You will be asked to take the science you have learned in this case and design a process or product to address a real-world issue of your choosing.

Engineers use the engineering design process as steps to address a real-world problem (see Figure 15.1). You will now use this process as you come up with a new way to use coil springs. In this case, you are asking the question (Step 1) of how you can design a new use for coil springs. Drawing on your creativity, you will then brainstorm (Step 2) a new product that uses coil springs to solve a problem. Afterward, you will create a plan (Step 3) for this new product. Next, you will create a sketch and/or model of your product (Step 4). Then, you will work with your classmates to think about how you would test (Step 5) and refine (Step 6) your product.

1. Ask Questions

Based on your previous research, consider a new problem that may be addressed or a product that could be created by using coil springs. What are some applications where you would need to use a coil to reduce or store energy? Are there any medical issues you can think of that could be solved with coil springs?

2. Brainstorm and Imagine

Springs come in many varieties and sizes. This allows researchers and engineers to use the items in a range of products, from medical devices to electronics to vehicles. They may also be used in specialized instruments in boats, aircraft, and even space rockets! Read the following blog entry about different types of springs and their current applications: *www.fictiv.com/hwg/design/types-of-springs-and-their-applications-an-overview*. Based on this information and your previous knowledge, brainstorm a specific product you could make using coil springs. (For example, a new type of coil spring could be added to shoes to reduce the impact energy of people with knee injuries). Remember that your product should be designed to solve a problem.

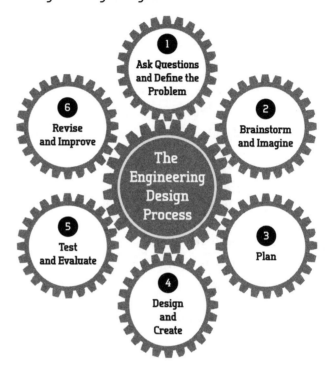

FIGURE 15.1

The Engineering Design Process

3. Create a Plan

Create a plan for your product. Consider: (1) What is the purpose of your product? (2) What are the benefits of your product? (3) What are the limitations of your product? Use the Product Planning Graphic Organizer (p. 245) to help you.

4. Design and Create

Consider the following questions and considerations for your coil springs product and its design.

- How would incorporating the coil spring in your design make the product better?

- Are there any limitations or drawbacks to using coil springs in your design? If so, how would you overcome them?

- What technologies might need to be developed to create or manufacture this design?

- What are any constraints or drawbacks you can foresee with implementing this design?

- Would there be any safety concerns regarding your coil springs–based product?

Create a sketch of your coil spring product design. Make sure your design incorporates your previous research and exploration.

5. Test and Evaluate

Working with your classmates, come up with a way to test your design to see its effectiveness.

6. Revise and Improve

Give your plans to one of your classmates for review. Listen to his or her feedback on your design. What are some ways you can use the input to refine your design? Take some time to revise and make improvements.

Reflect

1. What technologies might need to be developed to create or manufacture this design?

2. What are any constraints or drawbacks you can foresee with implementing this design?

3. Would there be any environmental or human health concerns about using this product?

Product Planning Graphic Organizer

Proposed Product Idea	
Pros (Benefits)	**Cons (Limitations)**

FROM SHIP TO STAIRCASE

THE HISTORY OF THE SLINKY

A Case Study Using the Discovery Engineering Process

Lesson Overview

In this lesson, students learn about the accidental invention of the Slinky, which is a coil spring (also called a helical spring). They review different uses for coil springs and come up with products that incorporate coil springs to solve a problem.

Lesson Objectives

By the end of this case study, students will be able to

- Describe the characteristics of coil springs.

- Analyze the benefits and limitations of coil springs as a device to store energy.

- Design a new application for coil springs that solves a current problem.

The Case Study Approach

This lesson uses a case study approach. Explaining the purpose of case studies will encourage your students to relate to the material and engage with the problem. At the heart of each case study in this book is a true story, one that describes how someone in his or her everyday life or during a routine workday made an observation or did a simple experiment that led to a new insight or discovery. Case studies are designed to get students actively engaged in the process of problem solving. The narrative of the case supplies authentic details that place the student in the role of the inventor and provide scaffolds for critical thinking and deep reflection. A case is more than a paragraph to read or a story to analyze but rather a way of framing problems, synthesizing what is known, and thinking creatively about new applications and solutions. In this lesson, students consider how the Slinky was discovered and work together to think about new applications for coil springs to solve real-life problems.

Use of the Case

Due to the nature of these case studies, teachers may elect to use any section of each case for their instructional needs. The sections are sequenced in order (scaffolded) so students think more deeply about the science involved in the case and develop an understanding of engineering in the context of science.

Curriculum Connections

Lesson Integration

You could use this case as a way to integrate engineering into a lesson on potential and kinetic energy, simple machines (incline plane), and waves (sound, light, and seismic waves).

Related Next Generation Science Standards

DISCIPLINARY CORE IDEA

- PS3.A: Definitions of Energy

  - Motion energy is properly called kinetic energy; it is proportional to the mass of the moving object and grows with the square of its speed. (MS-PS3-1)

  - A system of objects may also contain stored (potential) energy, depending on their relative positions. (MS-PS3-2)

SCIENCE AND ENGINEERING PRACTICES

- Analyzing and Interpreting Data
- Engaging in Argument From Evidence
- Constructing Explanations and Designing Solutions

CROSSCUTTING CONCEPTS

- Structure and Function
- Systems and System Models
- Energy and Matter

Related National Academy of Engineering Grand Challenge

- Engineer the Tools of Scientific Discovery

Lesson Preparation

You will need to make copies of the entire student section for the class. Students will need internet access at various points in the lesson. Alternatively, you can project videos or print and distribute copies of online content. Look at the Teaching Organizer (Table 15.1) for suggestions on how to organize the lesson.

Materials

For teacher demonstration:

- Toy coil spring
- Inclined plane or set of stairs

For each group of students:

- Toy coil spring
- Inclined plane or set of stairs
- Paperclips (1 per student)
- Pencils or other cylindrical objects (1 per student)
- Safety glasses or goggles (1 pair per student)

Safety Note for Students: Use caution not to trip over the coil spring on the floor. Wash your hands with soap and water immediately after completing this activity.

Time Needed

Up to 135 minutes

TABLE 15.1

Teaching Organizer

Section	Time Suggested	Materials Needed	Additional Considerations
The Case	10 minutes	Student packet	Could be done in class or as homework prior to class
Investigate (Coil Spring Demonstration)	10 minutes	Student packet, toy coil spring for teacher demonstration, inclined plane or set of stairs for teacher demonstration	Recommended as a whole-class activity
Investigate (Create, Innovate, and Investigate)	30–40 minutes	Student packet, toy coil spring, inclined plane or set of stairs, paperclips, pencils or other cylindrical objects, safety glasses or goggles	Recommended as a small-group activity
Apply and Analyze	10–15 minutes	Student packet, internet access	Individual activity
Design Challenge	45–60 minutes	Student packet, internet access	Small-group activity

Teacher Background Information

You may wish to review the concepts of simple machines (e.g., levers, pulleys, inclined planes, wedges, or screws that change the direction or magnitude of a force), how energy is transformed (e.g., potential and kinetic energy properties), and waves (e.g., sound, light, and seismic waves).

Vocabulary

- coil spring
- energy
- helical spring
- kinetic energy
- potential energy

Teacher Answer Key

Recognize, Recall, and Reflect

1. **What was Richard James's job during World War II?**

   James was a mechanical engineer who was trying to come up with a way to protect sensitive instruments in ships sailing through turbulent storms.

2. **Why are coil springs effective at keeping sensitive instruments in ships safe?**

   Coil springs absorb kinetic energy produced by the motion of the ship.

3. **Why do you believe the Slinky was not a toy that most stores initially wanted to sell? Why do you think the demonstration of the Slinky helped it become a favorite children's toy?**

   Answers will vary.

Questions for Reflection

1. **What did you observe about the movement and properties of a coil spring?**

   Answers will vary, but students might describe the toy coil spring's bounciness, its flexibility, and its looseness compared to other types of springs.

2. **What is unique about the coil spring?**

   Answers will vary.

Apply and Analyze

1. **What is chronic obstructive pulmonary disease (COPD)?**

   COPD is a broad term used to describe lung diseases that cause the airways to stiffen and swell, which results in difficulty breathing. As many as 65 million people worldwide have COPD and often have to receive medical treatments, including invasive surgery to remove diseased lung tissue.

2. **How do coils help treat individuals who have COPD?**

   The coils are inserted into the lungs and open healthy airways so that individuals are able to improve breathing.

3. **What are the benefits of using coils to treat COPD?**

   The coils reduce the chance of having an invasive surgery to remove damaged lung tissue. According to the study, more coil patients than non-coil patients were able to walk farther and reported a higher quality of life.

4. **According to the article, what were the challenges or downsides of using coils during the study? Why should researchers test the coils on a study group before allowing all doctors to use coils in their patients?**

   The article stated that more coil patients got pneumonia than non-coil patients. (Moreover, four participants in the coil group died compared to three in the non-coil group.) Testing treatments on small groups allows researchers to gauge how well they work and make improvements so that they will be even more effective. The tests also reveal what kind of side effects the treatments will have. Finally, testing keeps the general public safe, because a treatment or device will not be made available if it is deemed harmful.

Reflect

1. **What technologies might need to be developed to create or manufacture this design?**

   Answers will vary but could include new coil spring metals that are able to absorb or hold more energy for the design.

2. **What are any constraints or drawbacks you can foresee with implementing this design?**

   Any technologies developed would have to account for coil fatigue and the ability to hold energy.

3. **Would there be any environmental or human health concerns about using this product?**

   Any medical or environmental device would have to receive government approval (e.g., from the FDA). Other devices would need extensive testing to ensure safety for human/animal/environmental consumption.

Assessment

A rubric is included in the appendix (p. 377) that can be used to assess students' Design Challenge product.

Extensions

This lesson can be followed with lessons on energy transformation, properties of metals/alloys, and metal fatigue.

Resources and References

Associated Press. 2016. Implanted coils help some lung disease patients, study says. Fox News. *www.foxnews.com/health/2016/01/12/implanted-coils-help-some-lung-disease-patients-study-says.html*.

The Editors of Publications International, LTD. 9 things invented or discovered by accident. HowStuffWorks. *https://science.howstuffworks.com/innovation/scientific-experiments/9-things-invented-or-discovered-by-accident6.htm* (accessed December 17, 2017).

National Toy Hall of Fame. Slinky. The Strong/National Museum of Play. *www.toyhalloffame.org/toys/slinky* (accessed December 17, 2017).

Wu, Sylvia. 2017. Types of springs and their applications: An overview. *www.fictiv.com/hwg/design/types-of-springs-and-their-applications-an-overview* (accessed December 17, 2017).

16

MAN OF STAINLESS STEEL

The Discovery of a New Alloy

A Case Study Using the Engineering Design Process

Introduction

Humans have been creating products made from steel for thousands of years. Steel is made by adding carbon to iron. It is an example of an alloy, a material with the properties of a metal that is created by mixing more than one element together. The small amount of carbon in steel makes the metal harder, more ductile (capable of being drawn into threads), and stronger than iron alone. Yet, steel is vulnerable to rusting due to the large amounts of iron in it. Despite being seemingly indestructible, any substance that contains steel inevitably changes into a brittle, crumbly material that loses the beneficial properties of its original form. The creation of a new kind of alloy, stainless steel, was an accidental discovery that reshaped the world. With the physical strength of steel and a resistance to weathering and rust, stainless steel has helped propel society into modern times. The material has led to the invention of knives that do not corrode, safer surgical tools, and stronger automobiles.

Lesson Objectives

By the end of this case study, you will be able to

- Identify the chemical and physical properties of stainless steel.
- Differentiate between the properties of stainless steel and regular steel.

- Explain the conditions where rusting occurs.

- Design a product that uses stainless steel to solve a real-world problem.

The Case

Read the section below that describes the discovery of stainless steel. Then, answer the questions that follow.

Since its discovery, regular steel has been vulnerable to rusting. It wasn't until the early 20th century that metallurgists (scientists who are knowledgeable about metals) succeeded in their attempt to prevent steel from rusting by adding other elements to the alloy. The new formula was discovered in 1912 when metallurgist Harry Brearley made an accidental discovery while working to improve the design of a rifle.

The term *rifle* comes from the spiral grooves in a gun barrel that increase the accuracy of a shot by causing the bullet to spin. Repeated firing of a rifle, however, can make the barrel too wide for the bullet, because the friction from the bullet erodes the inside of the barrel. Brearley wanted to develop a steel alloy that could protect the inside of a gun barrel from the bullet. However, he found this task to be a more difficult than he had anticipated.

Brearley began diligently working to develop his new rifle barrel material, but his attempts were unsuccessful. His pile of failed steel scraps began to grow higher and higher in his laboratory. After months of failed attempts, Brearley noticed a shiny scrap of metal in his pile of rusting steel pieces. In addition to iron, the shiny sample of steel contained about 13% chromium and a small amount of carbon. A thin layer had formed on the surface of the steel as a result of the chromium reacting with oxygen in the air. This layer protected the steel from developing rust even if it was scratched.

While Brearley had not solved the problem he had originally set out to address, he did discover that his "rustless steel" resisted corrosion from vinegar and other acids such as lemon juice. Brearley's employers were not interested in using this new steel for the purpose of making rust-resistant rifles, but Brearley thought of another use for his discovery: as a material in knives.

In Brearley's time, most knives and other cutlery were made of steel, but keeping these utensils free of stains and rust required diligent care. Brearley thought his new alloy would be a perfect solution to this problem. He presented his manufacturing process to a local knife manufacturer. However, the manufacturer did not carefully follow Brearley's instructions and the knives that were produced in trial tests were not as hard as Brearley's samples, which made them difficult to grind and sharpen. The local townspeople joked that Brearley had invented knives that were too dull to cut. With additional experimentation, Brearley and his successors found the right combination of chromium and nickel that would keep the knives sharp.

Recognize, Recall, and Reflect

- What are the chemical ingredients in rustless steel?

- How does regular steel differ from stainless steel?

- Why did Brearley think stainless steel would be a good material with which to make cutlery?

Investigate

Corrosion is a type of chemical reaction in which a metal is attacked by an element in its environment and changed into an unwanted compound. The corrosion of iron, called rusting, is a familiar process. If you have ever spotted an old car or bicycle left outdoors for so long that it has rusted, you have seen the chemistry of corrosion in action!

Oxidation—when a chemical substance changes because of the introduction of oxygen—is one process that can lead to corrosion. When regular steel is exposed to environmental elements, it will eventually crumble through the process of oxidation. Regular steel is susceptible to oxidation (rusting), but stainless steel is not. Figure 16.1 shows photomicrographs (photos taken with the aid of a microscope) of regular steel (left) and stainless steel (right). You can see that the stainless steel is much smoother and less degraded than the regular steel.

FIGURE 16.1

Electron Photomicrographs of Regular Steel and Stainless Steel

 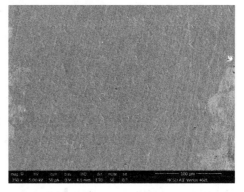

Stainless steel retains its stainless properties and does not rust due to the interaction between the elements in the steel alloy and the surrounding environment. Stainless steel contains iron, manganese, silicon, carbon, molybdenum, nickel, and most important, chromium. The elements that compose this alloy chemically react with oxygen and water in the atmosphere. This reaction produces a clear, thin film that is fairly stable, or nonreactive. Chromium in particular reacts with oxygen to produce a film that prevents water and oxygen from reaching the underlying layers that are susceptible to rusting. Chromium plays such an important role in this protective process that all stainless-steel alloys must be composed of at least 10% chromium.

In the next activity, you will watch a demonstration by your teacher or design an experiment to test how changes in an environment can affect the rate of rusting.

Materials

For groups of students or for a teacher demonstration:

- Regular steel wool or iron nail

- Water

- White vinegar

- Saltwater solution

- Test tubes

- Graduated cylinder

- Vegetable oil

- Boiled water

- Calcium chloride

- Rubber test tube stopper

- Indirectly vented chemical splash safety goggles (1 pair per student)

- Nonlatex apron (1 per student)

- Nitrile gloves (1 pair per student)

Safety Note: Wear indirectly vented chemical splash safety goggles, a nonlatex apron, and nitrile gloves during the setup, hands-on, and takedown segments of the activity. Use caution when working with hot liquids as they can cause skin burns. Use caution when using sharp objects as they can cut or puncture skin. Immediately clean up any liquid spilled on the floor, so it does not become a slip/fall hazard. Dispose of chemicals according to your teacher's instructions. Wash your hands with soap and water immediately after completing this activity.

Create, Innovate, and Investigate

Using the materials provided, design an experiment that would test to see which conditions are needed in order for an iron nail to rust. Here is some background information about your materials:

- Steel wool contains iron.

- Steel wool fibers usually come coated with oil so they do not rust. If you wanted the steel wool to rust, you could rinse the steel wool in vinegar.

- Saltwater and vinegar solutions both contain electrolytes. (Electrolytes are substances that dissociate into ions in solution.)

- Boiled water is made of H_2O.

- A layer of vegetable oil on top of boiled water prevents oxygen in the air from entering.

- Calcium chloride is a desiccant and absorbs any moisture (water) in the air.

Think of the following questions when designing the experiment:

1. How will you test to see if exposure to oxygen only (no water) can rust your steel?

2. How will you test to see if exposure to water only (no oxygen) can rust your steel?

3. How will you test to see if exposure to water and oxygen both can rust your steel?

4. How will you test to see if exposure to water, oxygen, and electrolytes can rust your steel?

Now, design a data table to collect your information, identifying your control and experimental groups in the table. Predict which conditions you think will cause the steel to rust the fastest. Make observations for 48 hours. Once you are done, create a graphic representation to compare the different solutions or conditions that you tested. How do you think you could use this information to solve problems of rusting automobiles? After you have completed this activity, answer the questions in the section that follows.

Questions for Reflection

1. Based on your observations, what conditions are necessary for rust to occur?

2. Which of the solutions produced the greatest amount of rust? Which solution produced the least amount of rust? What do you think caused this difference?

3. Rusting is faster when iron is exposed to electrolytes. Based on your observations, which of the solutions most likely contains the most electrolytes?

Apply and Analyze

Rusting occurs when iron reacts with oxygen in the presence of water. The iron passes electrons to the oxygen; the iron becomes more positively charged (this is called oxidation) and the oxygen becomes more negatively charged (this is called reduction). This reaction also creates a new compound called rust that is often reddish brown in color; the chemical name for rust is iron oxide (Fe_2O_3). Oxidation-reduction reactions are a very common chemical reaction.

1. What could you do if you wanted to prevent rusting in materials like iron or regular steel?

2. In which geographic regions do you think it's more likely that objects like cars will rust?

Design Challenge

Engineering is the application of scientific understanding through creativity, imagination, problem solving, and the designing and building of new materials to address and solve problems in the real world. You will be asked to take the science you have learned in this case and design a process or product to address a real-world issue of your choosing.

Engineers use the engineering design process as steps to address a real-world problem (see Figure 16.2). You will now use this process as you come up with a new way to use stainless steel. In this case, you are asking the question (Step 1) of how you can design a new use for

FIGURE 16.2

The Engineering Design Process

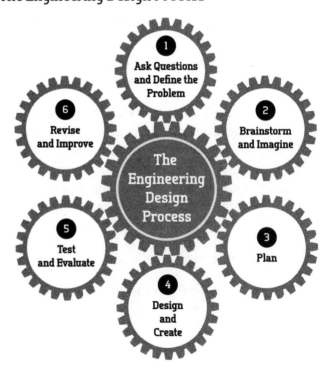

stainless steel. Drawing on your creativity, you will then brainstorm (Step 2) a new product that uses stainless steel to solve a problem. Afterward, you will create a plan (Step 3) for this new product. Next, you will create a sketch and/or model of your product (Step 4). Then, you will work with your classmates to think about how you would test (Step 5) and refine (Step 6) your product.

1. Ask Questions

Based on your previous research, consider a new problem that may be addressed or a product that could be created by using stainless steel. What are some applications where you would need a material that can maintain the strength of steel and withstand certain environmental conditions?

2. Brainstorm and Imagine

Brainstorm a specific product you could make using stainless steel. (As an example, you could use stainless steel to create a dog house that could keep pets sheltered and comfortable outdoors, while weathering the elements.) You may also think of an already existing product that uses regular steel that you would like to remake with stainless steel. (You should be able to explain why making this object out of stainless steel is more appropriate.)

3. Create a Plan

Create a plan for your product. Consider: (1) What is the purpose of your product? (2) What are the benefits of your product? (3) What are the limitations of your product? Use the Product Planning Graphic Organizer (p. 261) to help you.

4. Design and Create

Consider the following questions and considerations for your stainless steel product and its design.

- How would incorporating the stainless steel in your design make the product better?

- Are there any limitations or drawbacks to using stainless steel in your design? If so, how would you overcome them?

- What technologies might need to be developed to create or manufacture this design?

- What are any constraints or drawbacks you can foresee with implementing this design?

- Would there be any safety concerns regarding your stainless steel–based product?

Create a sketch of your stainless steel product design. Make sure your design incorporates your previous research and exploration.

5. Test and Evaluate

Working with your classmates, come up with a way to test your design to see its effectiveness.

6. Revise and Improve

Give your plans to one of your classmates for review. Listen to his or her feedback on your design. What are some ways you can use the input to refine your design? Take some time to revise and make improvements.

Reflect

1. What technologies might need to be developed to create or manufacture this design?

2. What are any constraints or drawbacks you can foresee with implementing this design?

3. Would there be any environmental or human health concerns about using this product?

Product Planning Graphic Organizer

Proposed Product Idea	
Pros (Benefits)	**Cons (Limitations)**

MAN OF STAINLESS STEEL

THE DISCOVERY OF A NEW ALLOY

A Case Study Using the Engineering Design Process

Lesson Overview

In this lesson, students will learn about the discovery of stainless steel. They will also analyze microscopic images of steel and stainless steel and design an experiment to determine the conditions needed for oxidation to take place. Finally, students will design a product that uses stainless steel to solve a real-world problem.

Lesson Objectives

By the end of this case study, students will be able to

- Identify the chemical and physical properties of stainless steel.

- Differentiate between the properties of stainless steel and regular steel.

- Explain the conditions where rusting occurs.

- Design a product that uses stainless steel to solve a real-world problem.

The Case Study Approach

This lesson uses a case study approach. Explaining the purpose of case studies will encourage your students to relate to the material and engage with the problem. At the heart of each case study in this book is a true story, one that describes how someone in his or her everyday life or during a routine workday made an observation or did a simple experiment that led to a new insight or discovery. Case studies are designed to get students actively engaged in the process of problem solving. The narrative of the case supplies authentic details that place the student in the role of the inventor and provide scaffolds for critical thinking and deep reflection. A case is more than a paragraph to read or a story to analyze but rather a way of framing problems, synthesizing what is known, and thinking creatively about new applications and solutions. In this lesson, students consider how stainless steel was discovered and work together to think of new applications for stainless steel to solve real-life problems.

Use of the Case

Due to the nature of these case studies, teachers may elect to use any section of each case for their instructional needs. The sections are sequenced in order (scaffolded) so students think more deeply about the science involved in the case and develop an understanding of engineering in the context of science.

Curriculum Connections

Lesson Integration

You could use this case as a way to integrate engineering into a lesson on chemical and physical properties, oxidation reactions, balancing chemical equations, and electrolytes.

Related Next Generation Science Standards

PERFORMANCE EXPECTATIONS

- MS-PS1-4. Develop a model that predicts and describes changes in particle motion, temperature, and state of a pure substance when thermal energy is added or removed.

- HS-ETS1-2. Design a solution to a complex real-world problem by breaking it down into smaller, more manageable problems that can be solved through engineering.

- HS-ETS1-3. Evaluate a solution to a complex real-world problem based on prioritized criteria and trade-offs that account for a range of constraints, including cost, safety, reliability, and aesthetics, as well as possible social, cultural, and environmental impacts.

SCIENCE AND ENGINEERING PRACTICES

- Analyzing and Interpreting Data
- Engaging in Argument From Evidence
- Constructing Explanations and Designing Solutions

CROSSCUTTING CONCEPT

- Structure and Function

Related National Academy of Engineering Grand Challenge

- Engineer the Tools of Scientific Discovery

Lesson Preparation

You will need to make copies of the entire student section for the class. Students will need internet access at various points in the lesson. Alternatively, you can project videos or print and distribute copies of online content. The Investigate portion of the lesson includes an inquiry into the conditions needed for oxidation (see the materials needed on p. 256/p. 265). If you don't have time for this activity, you may perform a demonstration for the students using the setup from Figure 16.3. Look at the Teaching Organizer (Table 16.1, p. 266) for tips on how to organize the lesson.

FIGURE 16.3

Demonstration Setup

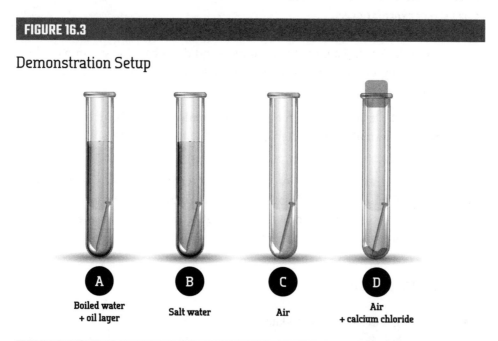

Test Tube	Conditions	Result
A	Boiled water + oil layer	No rust. Boiled water has no oxygen and oil stops new oxygen entering.
B	Salt water	Severe rust. Salt water is an electrolyte that conducts ions, speeding up rusting.
C	Air	Rust. Air and moisture cause normal rusting.
D	Air + calcium chloride	No rust. Calcium chloride dries out the air.

Materials

For each group of students or for teacher demonstration:

- Regular steel wool or iron nail

- Water

- White vinegar

- Saltwater solution

- Test tubes

- Graduated cylinder

- Vegetable oil

- Boiled water

- Calcium chloride

- Rubber test tube stopper

- Indirectly vented chemical splash safety goggles (1 pair per student)

- Nonlatex apron (1 per student)

- Nitrile gloves (1 pair per student)

Safety Note for Students: Wear indirectly vented chemical splash safety goggles, a nonlatex apron, and nitrile gloves during the setup, hands-on, and takedown segments of the activity. Use caution when working with hot liquids as they can cause skin burns. Use caution when using sharp objects as they can cut or puncture skin. Immediately clean up any liquid spilled on the floor, so it does not become a slip/fall hazard. Dispose of chemicals according to your teacher's instructions. Wash your hands with soap and water immediately after completing this activity.

Time Needed

3 class periods

TABLE 16.1

Teaching Organizer

Section	Time Suggested	Materials Needed	Additional Considerations
The Case	5 minutes	Student packet	Could be done in class or as homework prior to class
Investigate: Create, Innovate, Investigate	2 class periods	Student packet, regular steel wool or iron nail, water, white vinegar, saltwater solution, test tubes, graduated cylinder, vegetable oil, boiled water, calcium chloride, tubber test tube stopper, Indirectly vented chemical splash safety goggles, nonlatex apron, nitrile gloves	Small-group activity
Apply and Analyze	5 minutes	Student packet	Individual activity
Design Challenge	30 minutes	Student packet	Small-group activity

Teacher Background Information

In the Create, Innovate, and Investigate section, students may need assistance figuring out how to add the oil layer to the top of the water. They also may need additional explanation as to how the calcium chloride absorbs water from the air. (Calcium chloride [$CaCl_2$] is considered a hygroscopic solid, which means that the solid calcium chloride absorbs water from gases without dissolving in it.)

Vocabulary

- alloy
- desiccant
- dissociate

- electrolyte
- oxidation
- reduction

Teacher Answer Key

Recognize, Recall, and Reflect

1. **What are the chemical ingredients in rustless steel?**

   Rustless steel contains chromium, carbon, and iron (also manganese, silicon, molybdenum, and nickel). Chromium reacts with oxygen to produce a protective film.

2. **How does regular steel differ from stainless steel?**

    Regular steel typically does not have chromium and can rust when it comes in contact with oxygen and water.

3. **Why did Brearley think stainless steel would be a good material with which to make cutlery?**

    Brearley thought using stainless steel for cutlery would make it easier to keep cutlery free of stains and rust.

Questions for Reflection

1. **Based on your observations, what conditions are necessary for rust to occur?**

    Students should note that iron rusts in the presence of oxygen and water.

2. **Which of the solutions produced the greatest amount of rust? Which solution produced the least amount of rust? What do you think caused this difference?**

    The solution with salt water and the steel wool or nail will rust the most. The salt water acts like an electrolyte, speeding up the process.

3. **Rusting is faster when iron is exposed to electrolytes. Based on your observations, which of the solutions most likely contains the most electrolytes?**

    The solution with salt water.

Apply and Analyze

1. **What could you do if you wanted to prevent rusting in materials like iron or regular steel?**

    You could coat the material so that it is not exposed to water.

2. **In which geographic regions do you think it's more likely that objects like cars will rust?**

    Answers may vary. However, cars and other steel items typically develop rust more easily in coastal regions or wet environments.

Reflect

1. **What technologies might need to be developed to create or manufacture this design?**

    Answers will vary.

2.   **What are any constraints or drawbacks you can foresee with implementing this design?**

Answers will vary.

3.   **Would there be any environmental or human health concerns about using this product?**

Answers will vary.

Assessment

A rubric is included in the appendix (p. 377) that can be used to assess students' Design Challenge product.

Extensions

This lesson can be followed by discussions on how the loss of electrons from metal placed into a solution speeds the oxidation of that metal. The topic can also be addressed again during a comparison of reduction and oxidation reactions.

Resources and References

BBC Bitesize. Properties of metals. BBC. *www.bbc.com/bitesize/guides/zjb2pv4/revision/2.*

Brown, T. L., H. E. LeMay, B. E. Bursten, and J. R. Burdge. 2003. *Chemistry: The central science.* 9th ed. Upper Saddle River, NJ: Pearson Education, Inc.

Miodownik, M. 2014. *Stuff matters: Exploring the marvelous materials that shape our manmade world.* Boston: Houghton Mifflin Harcourt.

SUPER GLUE

Accidentally Discovered Twice

A Case Study Using the Discovery Engineering Process

Introduction

Humans have created glue for thousands of years from different sources, including animal tissues and tree bark. Recent advancements in chemistry have enabled humans

to develop new types of glues. Super Glue is a cyanoacrylate adhesive, a strong substance with special bonding properties that rapidly bonds, or glues, materials together. This glue has many household, cosmetic, and medical applications. It is used for everything from repairing shoes to attaching fake nail tips to fingernails during manicures. Doctors can even use the glue to repair skin or tissue torn in an accident. Read on to find out about the accidental discovery of Super Glue.

Lesson Objectives

By the end of the case study, you will be able to

* Describe the characteristics and uses of glue.

- Analyze the benefits and limitations of glue as a method to adhere materials together.

- Design a new application for glue that solves a current problem.

The Case

The following account outlines the accidental discovery of Super Glue and explains how the product emerged as one of the most popular glues for many different applications. Once you are finished reading, answer the questions that follow.

During World War II, Dr. Harry Coover and a team of scientists were looking for an adhesive that the military could use to bond gun pieces together. In 1942, they discovered a formula for a strong glue that rapidly adhered materials. Chemically, the glue formed durable chains of polymers (a substance composed of repeating subunits) when exposed to water. This allowed the glue to set or harden almost immediately. However, the glue was too sticky to be used with guns. So, Dr. Coover and his team decided to look for a different formula that better met their needs. The glue they had found, a type of cyanoacrylate substance, was soon forgotten.

In 1951, the substance was rediscovered. Dr. Coover and a colleague, Fred Joyner, were looking for a heat-resistant coating for jet cockpits. In experimenting with different substances, they once again came across the glue. Although they knew it wouldn't work as a cockpit coating, Coover and Joyner began to brainstorm other ideas for the product. They realized that it could be used for many different commercial applications, including electronics manufacturing and in the development of medical devices. Dr. Coover also documented the ability of the glue to be used as an adhesive for human tissue that had been cut. (In fact, during the Vietnam War it was used to treat soldiers as they waited for life-saving surgery.) Put on the market in 1958, the product eventually become known as Super Glue. Later, other similar glue products, generically called superglues, were marketed to consumers.

Recognize, Recall, and Reflect

1. What are the uses of Super Glue?

2. Why was Harry Coover researching and designing glue during World War II?

3. How was Super Glue discovered again?

Investigate

Imagine that you are a chemical engineer—an individual who uses chemistry to manufacture new substances or materials. You have been informed that your company needs a new glue that will hold two blocks together with additional weight added to the blocks. You will create glue from two different recipes and test each to determine which glue is the strongest.

Materials

For each group of students:

- 2 mixing bowls

- Spoons

- Clothespins

- Cardboard sheet or box

- Metric weights with hooks (1 g, 2 g, 5 g, and 10 g)

- 1 cup of flour (1st glue creation)

- ⅛ tsp. of salt (1st glue creation)

- ½ cup of water (1st glue creation)

- ½ cup of white flour (2nd glue creation)

- ⅓ cup of water (2nd glue creation)

- Heat source (hot plate) and pot (As an alternative, you may use a microwave and a Pyrex bowl.)

- Potholders

- Closed-toed shoes

- Indirectly vented chemical splash safety goggles (1 pair per student)

- Nonlatex apron (1 per student)

- Nitrile gloves (1 pair per student)

Safety Note: Wear indirectly vented chemical splash goggles, a nonlatex apron, and nitrile gloves during the setup, hands-on, and takedown segments of the activity. Use caution when working with hot liquids as they can cause skin burns. Use caution when working with hot plates as they can cause skin burns or electric shock. Immediately clean up any liquid or powder spilled on the floor so it does not become a slip/fall hazard. Dispose of chemicals according to your teacher's instructions. Wash your hands with soap and water immediately after completing this activity.

Create, Innovate, and Investigate

Read the directions described below and document your observations and data in the charts that follow.

GLUE CREATION #1 (NO HEAT REQUIRED)

- Gather your materials for Glue Creation #1 (mixing bowl, spoon, flour, water, salt; wear indirectly vented chemical splash safety goggles, nitrile gloves, closed-toed shoes, and a nonlatex apron).

  - Add the flour into the mixing bowl.

  - Slowly pour tiny amounts of water into the bowl to make a thick paste.

  - Add the salt to the paste and mix thoroughly with the spoon.

- Examine the glue you created. What do you observe (see, feel, and smell)? Record your observations in the Observation and Investigation Chart (Part 1) on page 274.

- Use your glue to attach a clothespin to a piece of cardboard. The glue will need to dry for 24 hours.

- Once the glue has dried, test the weight that your clothespin is able to hold while remaining attached to the cardboard. Clamp the clothespin on the hook of each metric weight while holding up the cardboard. (See Figure 17.1.) Test the 1-gram weight first, then 2 grams, 5 grams, and 10 grams. Record your observations in the Observation and Investigation Chart (Part 1). What did you notice about your glue?

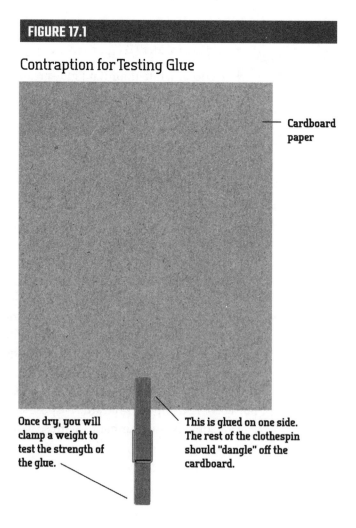

FIGURE 17.1

Contraption for Testing Glue

Cardboard paper

Once dry, you will clamp a weight to test the strength of the glue.

This is glued on one side. The rest of the clothespin should "dangle" off the cardboard.

GLUE CREATION #2 (HEAT REQUIRED)

- Gather your materials for Glue Creation #2 (mixing bowl, spoon, flour, water, heat source; wear indirectly vented chemical splash safety goggles, nitrile gloves, potholders [when placing the pot on or removing it from heat source], closed-toed shoes, and a nonlatex apron).

  - Add the flour to your mixing bowl.

  - Slowly pour tiny amounts of water into the bowl to make a thick paste.

  - Add the paste to a pot and place over medium heat (or use a Pyrex container and a microwave). **Use extra caution around the heat source.**

  - Stir the paste with your spoon until your paste begins to bubble. Once it begins to bubble, remove the pot from your heat source and turn off the heat source.

  - Let the paste cool.

- Examine the glue you created. What do you observe (see, feel, and smell)? Record your observations in the Observation and Investigation Chart (Part 2) on page 275.

- Glue a clothespin to a piece of cardboard. The glue will need to dry for 24 hours.

- Once the glue has dried, test the weight that your clothespin is able to hold while remaining attached to the cardboard. Clamp the clothespin on the hook of each metric weight while holding up the cardboard. Test the 1-gram weight first, then 2 grams, 5 grams, and 10 grams. Record your observations in the Observation and Investigation Chart (Part 2). What did you notice about your glue?

Observation and Investigation Chart (Part 1)

Glue Creation #1
What do you observe (see, feel, and smell)? Record your observations here.

Weight Test Observations			
1 Gram – Observation Notes	2 Grams – Observation Notes	5 Grams – Observation Notes	10 Grams – Observation Notes

What did you notice about Glue Creation #1?

Observation and Investigation Chart (Part 2)

Glue Creation #2

What do you observe (see, feel, and smell)? Record your observations here.

Weight Test Observations

1 Gram – Observation Notes	2 Grams – Observation Notes	5 Grams – Observation Notes	10 Grams – Observation Notes

What did you notice about Glue Creation #2?

Questions for Reflection

1. What were the differences between Glue Creation #1 and Glue Creation #2?

2. Which glue was stronger? Why do you think this glue recipe was stronger than the other glue?

3. If you were to design a new glue recipe, how would you change it to make the glue stronger so that the clothespin could hold more weight?

Apply and Analyze

Three-week-old Ashlyn Julian had a brain aneurysm (a bulge in a blood vessel that can cause internal bleeding, which may lead to death). Her doctors needed to find a way to save her without major surgery. Read the following story about how a type of superglue helped treat Ashlyn's aneurysm: *www.cbsnews.com/news/kansas-infants-brain-aneurysm-fixed-with-superglue.* Then answer the questions below.

1. What is an aneurysm? Why were doctors nervous to operate on the baby?

2. How did the doctors treat the aneurysm using superglue?

3. What are potential challenges in using superglue to treat newborn babies?

Design Challenge

Engineering is the application of scientific understanding through creativity, imagination, problem solving, and the designing and building of new materials to address and solve problems in the real world. You will be asked to take the science you have learned in this case and design a process or product to address a real-world issue of your choosing.

Engineers use the engineering design process as steps to address a real-world problem (see Figure 17.2). You will now use this process as you come up with a new way to use adhesive materials such as superglue. In this case, you are asking the question (Step 1) of how you can design a new use for an adhesive substance like superglue. Drawing on your creativity, you will then brainstorm (Step 2) a new product that uses an adhesive like superglue to solve a problem. Afterward, you will create a plan (Step 3) for this new product. Next, you will create a sketch and/or model of your product (Step 4). Then, you will work with your classmates to think about how you would test (Step 5) and refine (Step 6) your product.

1. Ask Questions

Based on your previous research, consider a new problem that may be addressed or a product that could be created by using a strong adhesive. What are applications where you would need a glue that could quickly and firmly bond two materials together? What problems could be solved with such a glue?

2. Brainstorm and Imagine

Different types of glue can be used in the creation of medical devices, electronics, and even buildings that can withstand sharply different temperatures over the course of the year (e.g., extreme hot or cold weather). In 2015, a superglue was discovered that is primarily made of water and can possibly be used as a protective coating for underwater surfaces such as submarines. (Read more about the glue here: *http://news.mit.edu/2015/hydrogel-superglue-water-adhesive-1109.*) Think about all of these applications, and then brainstorm your own idea for a new glue product. (For instance, one idea is to use glue to help bond broken pieces of a satellite in space.)

FIGURE 17.2

The Engineering Design Process

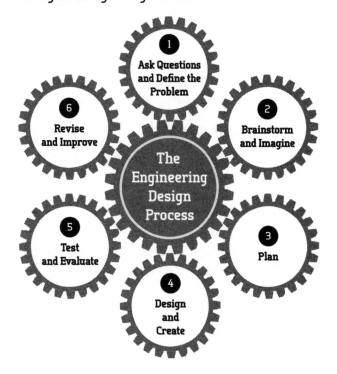

3. Create a Plan

Create a plan for your product. Consider: (1) What is the purpose of the glue? (2) What are the benefits of your product? (3) What are the limitations or drawbacks of your product? Use the Product Planning Graphic Organizer (p. 279) to help you develop your plan.

4. Design and Create

Consider the following questions and considerations for your glue product and its design.

- How would incorporating a strong glue into your design make the product better?

- Are there any limitations or drawbacks to using glue in your design? If so, how would you overcome them?

- What technologies might need to be developed to create or manufacture this design?

- What are any constraints or drawbacks you can foresee with implementing this design?

- Would there be any safety concerns regarding your product?

Create a sketch of your product design. Make sure your design incorporates your previous research and exploration.

5. Test and Evaluate

Working with your classmates, come up with a way to test your design to see its effectiveness.

6. Revise and Improve

Give your plans to one of your classmates for review. Listen to his or her feedback on your design. What are some ways you can use the input to refine your design? Take some time to revise and make improvements.

Reflect

1. What technologies might need to be developed to create or manufacture this design?

2. What are any constraints or drawbacks you can foresee with implementing this design?

3. Would there be any environmental or human health concerns about using this product?

Product Planning Graphic Organizer

Proposed Product Idea	
Pros (Benefits)	**Cons (Limitations)**

SUPER GLUE

ACCIDENTALLY DISCOVERED TWICE

A Case Study Using the Discovery Engineering Process

Lesson Overview

In this lesson, students explore the history of Super Glue and the applications of glues that can quickly bond two or more materials together with different environmental ranges. The development and application of Super Glue was an accidental discovery (twice realized) that has revolutionized manufacturing.

Lesson Objectives

By the end of this case study, students will be able to

- Describe the characteristics and uses of glue.

- Analyze the benefits and limitations of glue as a method to adhere materials together.

- Design a new application for glue that solves a current problem.

The Case Study Approach

This lesson uses a case study approach. Explaining the purpose of case studies will encourage your students to relate to the material and engage with the problem. At the heart of each case study in this book is a true story, one that describes how someone in his or her everyday life or during a routine workday made an observation or did a simple experiment that led to a new insight or discovery. Case studies are designed to get students actively engaged in the process of problem solving. The narrative of the case supplies authentic details that place the student in the role of the inventor and provide scaffolds for critical thinking and deep reflection. A case is more than a paragraph to read or a story to analyze but rather a way of framing problems, synthesizing what is known, and thinking creatively about new applications and solutions. In this lesson, students consider how Super Glue was discovered and work together to think about new applications for adhesives to solve real-life problems.

Use of the Case

Due to the nature of these case studies, teachers may elect to use any section of each case for their instructional needs. The sections are sequenced in order (scaffolded) so students think more deeply about the science involved in the case and develop an understanding of engineering in the context of science.

Curriculum Connections

Lesson Integration

You could use this case as a way to integrate engineering into a lesson on chemistry involving bonding, polymers, and how the environment affects properties of materials.

Related Next Generation Science Standards

PERFORMANCE EXPECTATIONS

- MS-PS1-2. Analyze and interpret data on the properties of substances before and after the substances interact to determine if a chemical reaction has occurred.

- MS-PS1-3. Gather and make sense of information to describe that synthetic materials come from natural resources and impact society.

SCIENCE AND ENGINEERING PRACTICES

- Analyzing and Interpreting Data

- Engaging in Argument From Evidence

- Constructing Explanations and Designing Solutions

CROSSCUTTING CONCEPT

- Structure and Function

Related National Academy of Engineering Grand Challenge

- Engineer the Tools of Scientific Discovery

Lesson Preparation

You will need to make copies of the entire student section for the class. Students will need internet access at various points in the lesson. Alternatively, you can project videos or print and distribute copies of online content. Look at the Teaching Organizer (Table 17.1) for suggestions on how to organize the lesson.

Materials

For each group of students:

- 2 mixing bowls

- Spoons

- Clothespins

- Cardboard sheet or box

- Metric weights with hooks (1 g, 2 g, 5 g, and 10 g)

- 1 cup of flour (1st glue creation)

- ⅛ tsp. of salt (1st glue creation)

- ½ cup of water (1st glue creation)

- ½ cup of white flour (2nd glue creation)

- ⅓ cup of water (2nd glue creation)

- Heat source (hot plate) and pot (As an alternative, you may use a microwave and a Pyrex bowl.)

- Potholders

- Closed-toed shoes

- Indirectly vented chemical splash safety goggles (1 pair per student)

- Nonlatex apron (1 per student)

- Nitrile gloves (1 pair per student)

Safety Note for Students: Wear indirectly vented chemical splash goggles, a nonlatex apron, and nitrile gloves during the setup, hands-on, and takedown segments of the activity. Use caution when working with hot liquids as they can cause skin burns. Use caution when working with hot plates as they can cause skin burns or electric shock. Immediately clean up any liquid or powder spilled on the floor so it does not become a slip/fall hazard. Dispose of chemicals according to your teacher's instructions. Wash your hands with soap and water immediately after completing this activity.

Time Needed

3–4 class periods

TABLE 17.1

Teaching Organizer

Section	Time Suggested	Materials Needed	Additional Considerations
The Case	10 minutes	Student packet	Could be read in class or as a homework assignment prior to class
Investigate	2 class periods	Student packet, materials to make glue (see list)	Recommended as a small-group activity
Apply and Analyze	10–15 minutes	Student packet, internet access	Small-group or individual activity where students compare results with peers
Design Challenge	45 minutes	Student packet, internet access	Small-group activity

Teacher Background Information

You may wish to review the following concepts:

- Bonding (e.g., the chemical bonds involved in creating material such as glue)

- Polymers (created by the bonding of monomers)

- How environmental factors (e.g., changes in temperature) affect the properties of materials

Vocabulary

- adhesive
- aneurysm
- cyanoacrylate
- polymer

Teacher Answer Key

Recognize, Recall, and Reflect

1. **What are the uses of Super Glue?**

    Super Glue can be used to bond materials together in the creation of military equipment, furniture, cosmetics, medical devices, electronic devices, and many other substances. It can also be used during medical procedures to bond organs/tissues.

2. Why was Harry Coover researching and designing glue during World War II?

He wanted to find a way to bond gun parts together as part of the World War II effort.

3. How was Super Glue discovered again?

After Dr. Coover and his team determined that the glue was not appropriate for bonding gun parts, the substance was forgotten. In the 1950s, Coover and colleague Fred Joyner came across the glue again while searching for a coating for jet cockpits and determined that there were many commercial uses for the product.

Questions for Reflection

1. What were the differences between Glue Creation #1 and Glue Creation #2?

Student answers will vary, but differences can be attributed to the use of a heat source in one recipe and not the other and the different materials used to create each glue.

2. Which glue was stronger? Why do you think this glue recipe was stronger than the other glue?

Student answers will vary based on data collection.

3. If you were to design a new glue recipe, how would you change it to make the glue stronger so that the clothespin could hold more weight?

Student answers will vary. But students may discuss making changes to the materials used, the type of heat source used, and the amount time spent preparing each glue.

Apply and Analyze

1. What is an aneurysm? Why were doctors nervous to operate on the baby?

An aneurysm is a bulge or bubble created within a blood vessel that can break, resulting in significant bleeding and death. The typical surgery required to address the problem often does not work well in very young children, especially infants.

2. How did the doctors treat the aneurysm using superglue?

Once the doctors removed the damaged area of the baby's blood vessel, they used the superglue to reattach the tissue.

3. What are potential challenges in using superglue to treat newborn babies?

Answers will vary. Challenges could include allergies to glue components, an immune system reacting to the glue, and dislocation of the glue from the original area.

Reflect

1. **What technologies might need to be developed to create or manufacture this design?**

    Answers will vary depending on the students' products.

2. **What are any constraints or drawbacks you can foresee with implementing this design?**

    Answers will vary; however, any technologies developed would have to account for adhesion properties in diverse or extreme environments.

3. **Would there be any environmental or human health concerns about using this product?**

    Answers will vary. However, any medical or environmental device would have to have government approval. Glues would need extensive testing to ensure safety for humans/animals/the environment.

Assessment

The Design Challenge can be assessed using the rubric in the appendix (p. 377).

Extensions

This lesson can be followed with lessons on chemistry-related properties of matter, adhesion, bonding, environmental effects on chemicals, and energy transformation.

Resources and References

Castillo, M. 2013. Kansas infant's brain aneurysm fixed with superglue. CBS News. *www.cbsnews.com/news/kansas-infants-brain-aneurysm-fixed-with-superglue.*

CEDESA Adhesive Solutions. The invention of Super Glue. *www.cedesa.co.uk/who-invented-superglue.html* (accessed December 17, 2017).

Chu, J. 2015. Hydrogel superglue is 90 percent water. MIT. *http://news.mit.edu/2015/hydrogel-superglue-water-adhesive-1109.*

Harris, E. *New York Times.* 2011. Harry Coover, Super Glue's inventor, dies at 94. March 27.

The Original Super Glue Corporation. History of Super Glue. *http://www.supergluecorp.com/?q=history.html* (accessed December 17, 2017).

TEFLON

A Refrigerator Gas Accident

A Case Study Using the Discovery
Engineering Process

Introduction

Teflon is best known as the material used as a nonstick coating for pots and pans so that food does not bond or attach to the cookware (see Figure 18.1). Additionally, Teflon coating is used in containers that may hold corrosive (destructive, eroding) chemicals that could dissolve human tissue. The coating is also found on certain medical tools, protecting humans by hindering or restricting the ability of infectious bacteria to stick to the medical instruments when the tools are inserted into human bodies.

FIGURE 18.1

Pan Made With Teflon

Teflon is a fluorocarbon polymer. Polymers are molecules consisting of repeating subunits (called monomers). Fluorocarbon is a molecule made of fluorine and carbon atoms. (Atoms are the smallest unit of matter consisting of protons, neutrons, and electrons). Fluorocarbons are very stable molecules. They are used to make plastics, refrigerants, and lubricants. Teflon is also hydrophobic, which means that it is not dissolved by water and appears to be repelled, or not attracted, to water. Teflon also has a high heat resistance, which means the molecules can handle a lot of heat. These properties combine to give Teflon its nonstick properties.

Lesson Objectives

By the end of the case study, you will be able to

- Describe the characteristics of Teflon.

- Review states of matter, phase changes of matter, and evaporative cooling.

- Investigate the molecular structure of plastic polymers like Teflon.

- Design a new application for plastic polymers that solves a current problem.

The Case

The following account outlines the accidental discovery of Teflon and explains how Teflon has been transformed into one of the most popular nonstick coatings for commercial use. Once you are finished reading the account, answer the questions that follow.

Dr. Roy Plunkett was a chemist who was designing a new refrigerant. Refrigerants are chemicals used in refrigerators and air conditioners for cooling. They support the process of evaporative cooling. In this process, refrigerants remove heat from an area being cooled and release the heat away from the cooling area. As part of Plunkett's study in 1938, he tried to create a new refrigerant from a gas called tetrafluoroethylene. When Plunkett opened one of the cylinders of gas that had been frozen earlier, he noticed that nothing came out of the cylinder. He cut open the cylinder and found that the tetrafluoroethylene had turned into a white material that was waxy and slippery. He named this new material Teflon. (The generic name for the chemical is polytetrafluoroethylene.) Due to its properties, there are many applications for Teflon (e.g., as a nonstick coating for pots and pans, a coating for bicycles that protects gears from moisture, a furniture coating that protects the upholstery from stains, a coating for medical instruments). But there are also environmental concerns about manufacturing and using plastics such as Teflon, including environmental contamination, the inability to recycle the material, and ozone depletion.

Recognize, Recall, and Reflect

1. How did Roy Plunkett's investigation of refrigerants lead to the discovery of Teflon?

2. What are the applications of Teflon?

Dive Deeper: States of Matter and Phase Changes

You just read that Teflon was created after matter being used by Dr. Roy Plunkett changed states (from gas to solid) through a phase change while Plunkett was looking into evaporative cooling. To learn more about states of matter, phase changes, and evaporative cooling, complete the activities below.

Activity 1: States of Matter, Water

In this activity, you will use water to review the states of matter. Record your information in the States of Matter Chart (p. 290).

MATERIALS

For each group of students:

- 100 ml beaker

- Ice cube

- Stopwatch

- Safety goggles (1 pair per student)

Safety Note: Wear safety goggles during the setup, hands-on, and takedown segments of the activity. Immediately wipe up any spilled water on the floor as it is a slip/fall hazard.

DIRECTIONS

- Place an ice cube in the 100 ml beaker. Record your observation and information about the ice cube in the first row of the chart.

- Leave the beaker with the ice cube out for 15 minutes. Once 15 minutes has passed, record your observation and information about what happened to the ice cube in the second row of the chart.

- What do you think would happen if you left the beaker out for a week? Record your prediction in the third row.

States of Matter Chart

State of Matter	**Written Description** Describe how the water looks in the beaker.	**Observation** Draw a picture of what you see happening in the beaker.	**Diagram of Water Molecules** Draw a picture of what you think the water molecules look like. Are they close or separated?
Solid			
Liquid			
Prediction: What do you think would happen if you left the beaker out for a week?			

Activity 2: Phase Changes

In this activity, you will place each term listed below in the correct box within the Phase Changes Diagram.

Phase Changes Diagram

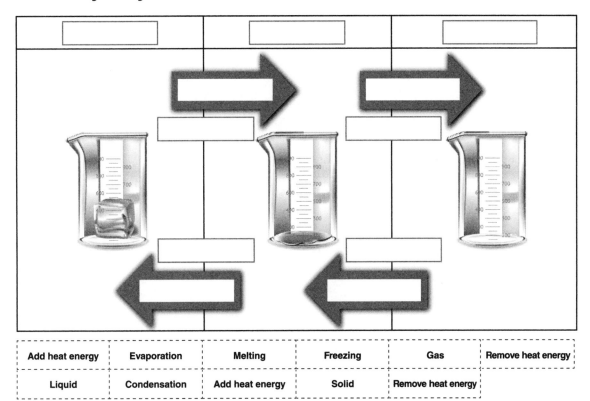

| Add heat energy | Evaporation | Melting | Freezing | Gas | Remove heat energy |
| Liquid | Condensation | Add heat energy | Solid | Remove heat energy | |

Activity 3: Evaporative Cooling Investigation

In this activity, you will investigate evaporative cooling. As shared in the case study, evaporative cooling helps cool areas by removing heat from the location that needs to be cold and releasing that heat outside of the cooling area. Follow the directions below, then answer the questions.

MATERIALS

For each group of students:

- Cosmetic cotton pad

- Rubbing alcohol

- Indirectly vented chemical splash safety goggles (1 pair per student)

- Nonlatex apron (1 per student)

- Nitrile gloves (1 pair per student)

Safety Note: Alert your teacher if you have alcohol or skin allergies. Wear indirectly vented chemical splash safety goggles, nitrile gloves, and a nonlatex apron during the setup, hands-on, and takedown segments of the activity. Alcohol is extremely flammable—keep it away from any active flame sources. Wash your hands with soap and water immediately after completing this activity.

DIRECTIONS

- Add rubbing alcohol to a cosmetic cotton pad.

- While wet, wipe the cosmetic cotton pad on your skin.

- Record your observations in the Evaporative Cooling Chart.

Evaporative Cooling Chart

What did you observe (see, feel) when you applied the rubbing alcohol to your skin?

What state of matter was the alcohol when it was first applied to your skin?	What state of matter was the alcohol after it was applied to your skin?

Draw a diagram or illustration of the phase changes that occurred with the alcohol.

What happened to the heat from your skin when the alcohol was applied?

The rubbing alcohol experiment provides an actual example of evaporative cooling. How does the rubbing alcohol show the way evaporative cooling works?

Investigate

Now that you have explored the processes behind the creation of Teflon, you will look into the molecular structure of the material. Teflon is a polymer plastic made of repeating subunits called monomers. You can create your own polymer by transforming milk into plastic! Milk contains a protein called casein. In this activity, you will link many casein molecules together to create your polymer.

Materials

For each group of students:

- Whole milk
- 2 disposable, heat-resistant cups (1 for each round)
- Food coloring
- Measuring cup
- Measuring spoons
- White vinegar
- Water
- Paper towels
- Bowl
- Tape or large rubber bands
- Plastic spoons
- Scale
- Heat source (stove or microwave)
- Pot/pan or Pyrex container
- Food coloring
- Zipper sandwich bag
- Indirectly vented chemical splash goggles (1 pair per student)
- Heat-resistant gloves (1 pair per student)
- Closed-toed shoes
- Nonlatex apron (1 per student)

Safety Note: The creation of polymer plastics will require a significant amount of heat. Extra caution will need to be exercised when dealing with hot materials and tools. Wear indirectly vented chemical splash safety goggles, heat-resistant gloves, and a nonlatex apron during the setup, hands-on, and takedown segments of the activity. Use caution when working with hot objects or heat sources as they can cause skin burns. Never eat any food items used in a lab activity. Wash your hands with soap and water immediately after completing this activity.

Create, Innovate, and Investigate

PART 1

Read the directions to create your polymer ball. Then document your observations and responses to questions in the Polymer Plastic Chart (p. 296).

- Pour 1 cup of milk into a pot/pan and heat it until it is bubbling. (Optional: Add a drop or two of food coloring to the milk based on your preference of color [you may mix colors].)

- In a heat-resistant cup, add 5 teaspoons of white vinegar.

- Add the hot milk into the heat-resistant cup. Use a spoon to mix the milk and vinegar together. Clusters of white objects called curds will begin to form.

- Add three paper towels (stacked one on top of the other) on top of a bowl and secure with tape or a rubber band. Pour the mixture from the heat-resistant cup onto the paper towel–covered bowl. (POUR SLOWLY.) The paper towels will filter the liquid from the curds. The bowl will catch the liquids and the paper towels will catch the curds.

- Scrape the curds from the towel and blot any excess liquid found on the curds using a paper towel.

- Weigh the collected curds. Record the weight (in grams) in the Polymer Plastic Chart.

- The curds you have are primarily casein. While the curds are still warm, mold and shape the casein. Once the casein polymer has dried (this will take at least 24 hours), seal it up in a zippered bag for safekeeping.

- Answer the questions in the Polymer Plastic Chart.

Polymer Plastic Chart

Record the weight (in grams) of the casein produced.
What did you notice about your polymer plastic? What did you see and feel?
What is the purpose of using white vinegar in this process? Research the answer online.
There are many variables or steps in the process that could change how much casein you can extract (or produce) from milk. Which factors might affect how much casein can be extracted from milk? How would you change the formula/directions above to produce more casein? Why would you make these changes?

PART 2

In this activity, you will re-create your polymer plastic based on your new formula outline above. Document your observations and responses to questions in the chart below.

In grams, how much casein was produced with your new formula?
What did you notice about your polymer plastic? What did you see and feel?
How was your new polymer plastic different from your first one? What changes in your formula directly affected the properties of your new polymer plastic?
Based on your observations, describe the phase changes throughout this process.

Apply and Analyze

Although Teflon has revolutionized products like medical tools and cookware, the chemicals used to create this and other plastic materials have been linked to environmental issues and health problems. Read this story about the detection of these chemicals in one region's water supply: *http://fortune.com/longform/teflon-pollution-north-carolina*. Then answer the questions below.

1. Why are individuals concerned about the water in North Carolina? What is causing the chemical pollution in the water?

2. What is the purpose of the Environmental Protection Agency (EPA)? Why is the EPA involved in this matter?

3. How might the chemicals in the water affect ecosystems and biomes? What might be the long-term effects on living organisms?

Design Challenge

Engineering is the application of scientific understanding through creativity, imagination, problem solving, and the designing and building of new materials to address and solve problems in the real world. You will be asked to take the science you have learned in this case and design a process or product to address a real-world issue of your choosing.

Engineers use the engineering design process as steps to address a real-world problem (see Figure 18.2). You will now use this process as you come up with a new way to use a Teflon-like plastic polymer. In this case, you are asking the question (Step 1) of how you can design a new use for plastic polymers. Drawing on your creativity, you will then brainstorm (Step 2) a new product that uses plastic polymers to solve a problem.

FIGURE 18.2

The Engineering Design Process

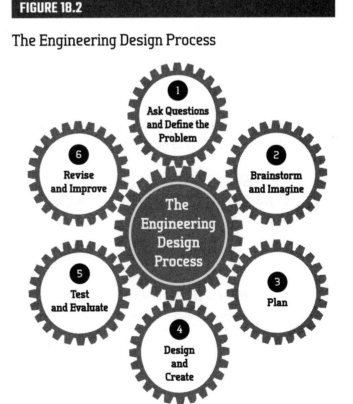

Afterward, you will create a plan (Step 3) for this new product. Next, you will create a sketch and/or model of your product (Step 4). Then, you will work with your classmates to think about how you would test (Step 5) and refine (Step 6) your product.

1. Ask Questions

Based on your previous research, consider a new problem that may be addressed or product that could be created by using plastic polymers. What are applications where you need a nonstick, protective material? What problem could be solved with plastic polymers?

2. Brainstorm and Imagine

Read the Application and Uses section of the following Wikipedia page on polytetrafluoroethylene (aka Teflon) to learn more about current uses for this plastic polymer: *https://en.wikipedia.org/wiki/Polytetrafluoroethylene#Applications_and_uses*. Next, brainstorm a new product that uses similar plastic polymers to solve a problem. (For example, maybe a new type of plastic polymer could be added to walls of spacecraft to protect people in space from radiation.) Think of ways in which you can make your product eco-friendly.

3. Create a Plan

Create a plan for your product. Consider: (1) What is the purpose of your plastic polymer–based product? (2) What are benefits of your product? (3) What are the limitations or drawbacks of your product? Use the Product Planning Graphic Organizer (p. 300) to help you develop your plan.

4. Design and Create

Consider the following questions and considerations for your product and its design.

- How would incorporating plastic polymers into your design make the product better?

- Are there any limitations or drawbacks to using plastic polymers in your design? If so, how would you overcome them?

- What technologies might need to be developed to create or manufacture this design?

- What are any constraints or drawbacks you can foresee with implementing this design?

- Would there be any safety concerns regarding your product?

Create a sketch of your product design. Make sure your design incorporates your previous research and exploration.

5. Test and Evaluate

Working with your classmates, come up with a way to test your design to see its effectiveness.

6. Revise and Improve

Give your plans to one of your classmates for review. Listen to his or her feedback on your design. What are some ways you can use the input to refine your design? Take some time to revise and make improvements.

Reflect

1. What technologies might need to be developed to create or manufacture this design?

2. What are any constraints or drawbacks you can foresee with implementing this design?

3. Would there be any environmental or human health concerns about this design?

Product Planning Graphic Organizer

Proposed Product Idea

Pros (Benefits)	Cons (Limitations)

TEFLON

A REFRIGERATOR GAS ACCIDENT

A Case Study Using the Discovery Engineering Process

Lesson Overview

In this lesson, students explore the accidental discovery of Teflon and learn about this plastic polymer's structure and uses. Students then review states of matter, phase changes, and evaporative cooling. They also create a polymer plastic from milk and discuss the environmental concerns regarding plastics. Finally, students design a new application for a plastic polymer that solves a current problem.

Lesson Objectives

By the end of this case study, students will be able to

- Describe the characteristics of Teflon.

- Review states of matter, phase changes of matter, and evaporative cooling.

- Investigate the molecular structure of plastic polymers like Teflon.

- Design a new application for plastic polymers that solves a current problem.

The Case Study Approach

This lesson uses a case study approach. Explaining the purpose of case studies will encourage your students to relate to the material and engage with the problem. At the heart of each case study in this book is a true story, one that describes how someone in his or her everyday life or during a routine workday made an observation or did a simple experiment that led to a new insight or discovery. Case studies are designed to get students actively engaged in the process of problem solving. The narrative of the case supplies authentic details that place the student in the role of the inventor and provide scaffolds for critical thinking and deep reflection. A case is more than a paragraph to read or a story to analyze but rather a way of framing problems, synthesizing what is known, and thinking creatively about new applications and solutions. In this lesson, students consider how the plastic polymer Teflon was discovered and work together to think about new applications for plastic polymers to solve real-life problems.

Use of the Case

Due to the nature of these case studies, teachers may elect to use any section of each case for their instructional needs. The sections are sequenced in order (scaffolded) so students think more deeply about the science involved in the case and develop understanding of engineering in the context of science.

Curriculum Connections

Lesson Integration

You could use this case as a way to integrate engineering into a lesson on phases of matter, phase changes of matter, polymers, chemical reactions, environmental impact of plastics, and physical and chemical properties of substances.

Related Next Generation Science Standards
DISCIPLINARY CORE IDEA

- PS1.A. Structure and Properties of Matter

  ○ Substances are made from different types of atoms, which combine with one another in various ways. Atoms form molecules that range in size from two to thousands of atoms. (MS-PS1-1)

  ○ Solids may be formed from molecules, or they may be extended structures with repeating subunits (e.g., crystals). (MS-PS1-1)

  ○ Each pure substance has characteristic physical and chemical properties (for any bulk quantity under given conditions) that can be used to identify it. (MS-PS1-2, MS-PS1-3)

  ○ Gases and liquids are made of molecules or inert atoms that are moving about relative to each other. (MS-PS1-4)

  ○ In a liquid, the molecules are constantly in contact with others; in a gas, they are widely spaced except when they happen to collide. In a solid, atoms are closely spaced and may vibrate in position but do not change relative locations. (MS-PS1-4)

  ○ The changes of state that occur with variations in temperature or pressure can be described and predicted using these models of matter. (MS-PS1-4)

SCIENCE AND ENGINEERING PRACTICES

- Analyzing and Interpreting Data
- Constructing Explanations and Designing Solutions

CROSSCUTTING CONCEPTS

- Structure and Function
- Stability and Change

Related National Academy of Engineering Grand Challenge

- Engineer the Tools of Scientific Discovery

Lesson Preparation

You will need to make copies of the entire student section for the class. In the Investigate section, students will be creating plastic polymers out of milk. Test the milk you plan to use in the activity beforehand. You may encounter problems with ultra-pasteurized milk. Alternatives to ultra-pasteurized milk are pasteurized milk or unpasteurized milk. Students will need internet access to read an article for the Apply and Analyze section. Alternatively, you can print and distribute copies of the article to students. Look at the Teaching Organizer (Table 18.1, p. 305) for suggestions on how to organize the lesson.

Materials

For each group of students:

DIVE DEEPER: ACTIVITY 1

- 100 ml beaker
- Ice cube
- Stopwatch
- Safety goggles (1 pair per student)

Safety Note for Students: Wear safety goggles during the setup, hands-on, and takedown segments of the activity. Immediately wipe up any spilled water on the floor as it is a slip/fall hazard.

DIVE DEEPER: ACTIVITY 3

- Cosmetic cotton pad
- Rubbing alcohol
- Indirectly vented chemical splash safety goggles (1 pair per student)
- Nonlatex apron (1 per student)
- Nitrile gloves (1 pair per student)

Safety Note for Students: Alert your teacher if you have alcohol or skin allergies. Wear indirectly vented chemical splash safety goggles, nitrile gloves, and a nonlatex apron during the setup, hands-on, and takedown segments of the activity. Alcohol is extremely flammable—keep it away from any active flame sources. Wash your hands with soap and water immediately after completing this activity.

INVESTIGATE: CREATE, INNOVATE, AND INVESTIGATE (PARTS 1 AND 2)

- Whole milk
- 2 disposable, heat-resistant cups (1 for each round)
- Food coloring
- Measuring cup
- Measuring spoons
- White vinegar
- Water
- Paper towels
- Bowl
- Tape or large rubber bands
- Plastic spoons
- Scale
- Heat source (stove or microwave)
- Pot/pan or Pyrex container
- Food coloring
- Zipper sandwich bag
- Indirectly vented chemical splash goggles (1 pair per student)

- Heat-resistant gloves (1 pair per student)

- Closed-toed shoes

- Nonlatex apron (1 per student)

Safety Note: The creation of polymer plastics will require a significant amount of heat. Extra caution will need to be exercised when dealing with hot materials and tools. Wear indirectly vented chemical splash safety goggles, heat-resistant gloves, and a nonlatex apron during the setup, hands-on, and takedown segments of the activity. Use caution when working with hot objects or heat sources as they can cause skin burns. Never eat any food items used in a lab activity. Wash your hands with soap and water immediately after completing this activity.

Time Needed

3–4 class periods

TABLE 18.1

Teaching Organizer

Section	Time Suggested	Materials Needed	Additional Considerations
The Case	10 minutes	Student packet	Could be read in class or as a homework assignment prior to class
Dive Deeper (Activities 1, 2, and 3)	40 minutes	Student packet; see material lists for Activities 1 and 3	Recommended as a whole-class activity
Investigate (Create, Innovate, and Investigate, Parts 1 and 2)	2 class periods	Student packet; see material list for Create, Innovate, and Investigate, Parts 1 and 2	Recommended as a small-group activity
Apply and Analyze	10–15 minutes	Student packet, internet access	Individual activity (Students can compare results with classmates.)
Design Challenge	45 minutes	Student packet	Small-group activity

Teacher Background Information

You may wish to review the concepts of phases of matter, phase changes of matter, polymers, chemical reactions, environmental impact of plastics, and physical and chemical properties of substances.

Vocabulary

- atoms
- bond
- casein
- converted
- corrosive
- erode
- evaporative cooling
- fluorocarbon

- hydrophobic
- monomer
- nonstick coating
- plastic
- plastic polymer
- polymer
- refrigerant
- Teflon

Teacher Answer Key

Recognize, Recall, and Reflect

1. **How did Roy Plunkett's investigation of refrigerants lead to the discovery of Teflon?**

   Plunkett tried to create a new refrigerant from a gas called tetrafluoroethylene. But when he opened one of the cylinders of gas that had been frozen earlier, he noticed that nothing came out. He cut open the cylinder and found that the tetrafluoroethylene had turned into a waxy and slippery material that would become Teflon.

2. **What are the applications of Teflon?**

   Answers will vary. But students might point out that it can be used as nonstick coating for pots and pans, a coating for bicycles that protects gears from moisture, a furniture coating that protects the upholstery from stains, etc.

Apply and Analyze

1. **Why are individuals concerned about the water in North Carolina? What is causing the chemical pollution in the water?**

   Chemours—the manufacturer that makes Teflon—released a chemical called GenX into the state's water supplies during its production of the material. People are concerned that the polluted water will harm animals and humans as GenX has been linked to certain kinds of cancer.

2. **What is the purpose of the Environmental Protection Agency? Why is the EPA involved in this matter?**

The EPA is a government agency that is designed to protect the health of the environment. When companies appear to be responsible for harmful, widespread pollution, as in the case of Chemours, the EPA carries out an investigation to determine the cause of the pollution and how to fix it.

3. **How might the chemicals in the water affect ecosystems and biomes? What might be the long-term effects on living organisms?**

   If dangerous chemicals are in the water, they could contaminate the organisms (plants, animals, and humans) that come into contact with them. The long-term effects of exposure to such chemicals aren't always clear. But it's possible they could include illness, deformities, and even death.

Reflect

1. **What technologies might need to be developed to create or manufacture this design?**

   Answers will vary but could include the need for new polymer plastics that are able to withstand high or low pressure and temperatures. Students will also want to think of ways to manufacture an eco-friendly product.

2. **What are any constraints or drawbacks you can foresee with implementing this design?**

   Answers will vary.

3. **Would there be any environmental or human health concerns about this design?**

   Students can cite the environmental and health concerns mentioned throughout the case study. They should think of ways their designs can address these concerns.

Assessment

The Design Challenge can be assessed using the rubric in the appendix (p. 377).

Extensions

This lesson can be followed with lessons on energy transformation, properties of metals/alloys, metal fatigue, and magnetics and creating magnets.

Resources and References

Kary, T., and J. Kaskey. 2017. On U.S. rivers, Teflon's old cancer ties are stoking new fears. Bloomberg. *www.bloomberg.com/news/articles/2017-07-06/on-u-s-rivers-teflon-s-old-cancer-ties-are-stoking-new-fears.*

Science Buddies. 2012. Sculpted science: Turn milk into plastic! Scientific American. *www.scientificamerican.com/article/bring-science-home-milk-plastic.*

Science History Institute. 2017. Roy J. Plunkett. *www.sciencehistory.org/historical-profile/roy-j-plunkett.*

Wikipedia. Polytetrafluoroethylene. *https://en.wikipedia.org/wiki/Polytetrafluoroethylene#Applications_and_uses* (accessed December 2017).

19

VASELINE

The "Wonder Jelly"

A Case Study Using the Discovery
Engineering Process

Introduction

Petroleum jelly, often sold under the brand Vaseline, is an inexpensive product that has been available to consumers for over a century. Over the years, it has been used for a wide range of things, including as a lure for catching fish, as lubricating grease, as a balm to heal chapped skin and small burns, and for cooking. It has even been used by NFL football players to stay warm in frigid temperatures. Often called a "wonder product" due to its versatility, there are many uses for petroleum jelly that have yet to be considered.

Lesson Objectives

By the end of this case study, you will be able to

- Describe how petroleum jelly was discovered.

- Analyze some of the useful properties of petroleum jelly.

- Design a new product that uses petroleum jelly to solve a problem.

The Case

Read the following story about the invention of Vaseline as summarized from the company's website and the book *Panati's Extraordinary Origins of Everyday Things* by Charles Panati. Once you are finished reading, answer the questions that follow.

In 1859, a 22-year-old chemist named Robert Augustus Chesebrough traveled to Titusville, Pennsylvania, where petroleum (crude oil) had just been discovered. Chesebrough wanted to research if any new products could be made from the fuel. While talking to workers who were drilling for the oil, Chesebrough learned that a waxlike material would form on their equipment as they drilled. This waxy substance was an annoyance to the workers, because it caused their machinery to malfunction. But the workers had also noticed something surprising about the material. They reported that it helped their cuts and burns heal faster, so they would rub it on their skin.

This gave Chesebrough an idea. He took what he learned from the oil workers in Titusville back to his lab in New York. He then worked for several years to create a process that would purify the waxy jelly from petroleum so that he could sell it to consumers. In 1865, about five years after his observations in Titusville, Chesebrough was ready to file a patent with the U.S. Patent Office. He was awarded U.S. patent US49230A, titled "Improvement in apparatus for filtering petroleum." As the patent application explained, Chesebrough's invention consisted of "the application of heat to a filter for hydrocarbon or other oils by means of three distinct cylinders or a steam-worm coiled inside or outside of the filter." In basic terms, Chesebrough had created a filtering device that used heat to refine the jelly. He would continue to improve his process for purifying petroleum jelly, eventually creating the product we know today as Vaseline.

In 1870, Chesebrough opened a factory in Brooklyn, New York, to manufacture his product. He also set to work on advertising the petroleum jelly and showing off its many uses. In front of a live audience, Chesebrough would burn his skin with a flame or acid and then demonstrate how to apply Vaseline to burns!

Eventually, Vaseline became popular with consumers (Figure 19.1). Many people use the product as skin balm. In addition to helping heal damaged skin, the

FIGURE 19.1

Vaseline From the Early 1900s

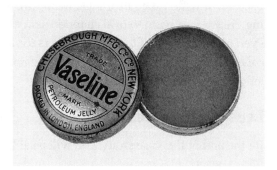

product is often used to protect skin and to keep in moisture. It is useful in cold environments, because it doesn't freeze easily. Commander Robert Peary even brought Vaseline with him to the North Pole during an expedition in 1909 to protect his skin from chapping and to keep his equipment from rusting in the extreme environment.

Recognize, Recall, and Reflect

1. Why did chemist Robert Chesebrough go to Titusville, Pennsylvania?

2. What did Chesebrough learn from the workers drilling for oil in Titusville?

3. Why was it necessary for Chesebrough to work in his lab for several years with the waxy petroleum jelly?

Investigate

In this activity, you will explore uses for petroleum jelly in cold temperatures.

Materials

For each group of students:

- Ice water

- Warm water

- Petroleum jelly

- 1000 ml beakers (2 per group)

- 1 graduated cylinder

- 4 zipper sandwich bags

- Plastic spoon

- 2 thermometers

- Scale to measure in grams

- Stopwatch or timing device

- Indirectly vented chemical splash safety goggles (1 pair per student)

- Nonlatex apron (1 per student)

- Nitrile gloves (1 pair per student)

Safety Note: Wear indirectly vented chemical splash safety goggles, a nonlatex apron, and nitrile gloves during the setup, hands-on, and takedown segments of the activity. Immediately wipe up any spilled water on the floor as it is a slip/fall hazard. Use caution when working with glassware, which can shatter if dropped and cut skin. Wash your hands with soap and water immediately after completing this activity.

Create, Innovate, and Investigate

- Create a data table with two columns. You will need to record the temperature once per minute for 10 minutes in each of the columns. Label one column "No Petroleum Jelly" and label the second column "Petroleum Jelly."

- Begin by adding 100 grams of petroleum jelly to the inside of a plastic sandwich bag (bag number 1). Use the spoon to spread the petroleum jelly around the inside of the bottom half of the bag. Place a second sandwich bag (bag number 2) inside the first. You should now have two sandwich bags separated by a thin layer of the jelly. This is your equipment for the "Petroleum Jelly" trial.

- Place the third sandwich bag (bag number 3) inside another clean sandwich bag (bag number 4). Do not add any petroleum jelly. This is your equipment for the "No Petroleum Jelly" trial.

- Pour 700 ml of ice water into each of the two large beakers.

- Use the graduated cylinder to measure and pour 150 ml of warm water into each of the two sets of plastic bags.

- Take the temperature of the water in each beaker and in each set of plastic bags before zipping the bags shut. Record your results. The starting temperatures of the two ice waters should be pretty similar to each other as should the starting temperatures of the two warm waters.

- Make a prediction. What will happen to the temperature of the water inside the two sets of bags when they are placed in cold water? Do you expect to see any changes in temperature? Write down your prediction next to your data table.

- Place the "No Petroleum Jelly" bags with warm water into one beaker and the "Petroleum Jelly" bags with warm water into the other beaker at the exact same time. Start your stopwatch.

- After one minute, record the temperature of the water inside the "No Petroleum Jelly" set of bags (your control) and record the temperature of the water inside the "Petroleum Jelly" set of bags (your experiment). Be sure not to let any of the cold water accidentally get inside the bag.

- Take the temperature inside the two sets of plastic bags every minute for 10 minutes, recording the data in your data table.

Questions for Reflection

1. Look at the data you collected. Did you notice any patterns?

2. Did your observations match your prediction? What was different?

3. What does the data you collected suggest about the properties of petroleum jelly?

4. Sometimes swimmers who swim in cold water smear petroleum jelly on their bodies. Based on your data, would petroleum jelly help keep them warm?

5. What remaining questions do you have about petroleum jelly?

6. How would you design an experiment to collect data related to your question?

Apply and Analyze

Petroleum jelly has often been called "wonder jelly" due to its many uses for skin care, but why is it effective? How does it work? Explore each resource listed below to learn more about petroleum jelly. Then answer the questions that follow.

- Article: "Can petroleum jelly be used as a moisturizer?" (*https://health. howstuffworks.com/skin-care/moisturizing/products/petroleum-jelly-moisturizer. htm/printable*)

- Video: "Mayo clinic minute: The many benefits of petroleum jelly" (*www. youtube.com/watch?v=ubpsosv7mHM*)

1. Describe what petroleum jelly does when rubbed onto the skin.

2. What are the benefits of using petroleum jelly for skin care?

3. Robert Chesebrough called petroleum jelly a wonder product and claimed that it accelerated healing. Based on the information presented by the article and the video, do you think his claim is accurate?

Design Challenge

Engineering is the application of scientific understanding through creativity, imagination, problem solving, and the designing and building of new materials to address and solve problems in the real world. You will be asked to take the science you have learned in this case and design a process or product to address a real-world issue of your choosing.

Engineers use the engineering design process as steps to address a real-world problem (see Figure 19.2). You will now use this process as you come up with a new way to use petroleum jelly. In this case, you are asking the question (Step 1) of how you can design a new use for petroleum jelly. Drawing on your creativity, you will then brainstorm (Step 2) a specific new product that uses petroleum jelly to solve a problem. Afterward, you will create a plan (Step 3) for this new product. Next, you will create a sketch and/or model of your product (Step 4). Then, you will work with your classmates to think about how you would test (Step 5) and refine (Step 6) your product.

1. Ask Questions

Based on your work in the sections above, consider a new problem that may be addressed or product that could be created using petroleum jelly. What are some applications where you would need a material that can trap moisture or form a protective barrier? How could petroleum jelly be used in new ways to help people with skin ailments?

2. Brainstorm and Imagine

Brainstorm a specific product that uses petroleum jelly to solve a problem. (For instance, you could invent a pair of gloves that dispenses the jelly to keep hands moisturized in cold weather.)

3. Create a Plan

Create a plan for your product. Consider: (1) What is the purpose of your product? (2) What are benefits to using petroleum jelly in your product? (3) What are the limitations of using petroleum jelly? Are there any problems you might anticipate? Use the Product Planning Graphic Organizer (p. 316) to help you.

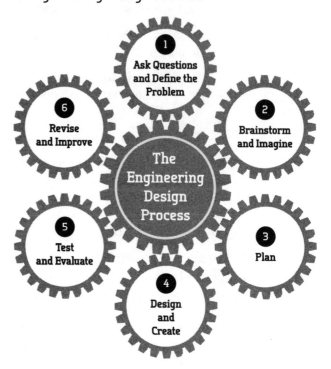

FIGURE 19.2

The Engineering Design Process

1. Ask Questions and Define the Problem
2. Brainstorm and Imagine
3. Plan
4. Design and Create
5. Test and Evaluate
6. Revise and Improve

The Engineering Design Process

4. Design and Create

Consider the following questions and considerations for your product and its design.

- How would incorporating petroleum jelly into your design make the product better?

- How would you overcome any limitations or drawbacks caused by using petroleum jelly in your design?

- What technologies might need to be developed to create or manufacture this design?

- What are any constraints or drawbacks you can foresee with implementing this design?

- Would there be any safety concerns regarding your product?

Create a sketch of your product design. Make sure your design incorporates your previous research and exploration.

5. Test and Evaluate

Working with your classmates, come up with a way to test your design to see its effectiveness.

6. Revise and Improve

Give your plans to one of your classmates for review. Listen to his or her feedback on your design. What are some ways you can use the input to refine your design? Take some time to revise and make improvements.

Reflect

1. What technologies might need to be developed to create or manufacture this design?

2. What are any constraints or drawbacks you can foresee with implementing this design?

3. Would there be any environmental or human health concerns about this design?

Product Planning Graphic Organizer

Proposed Product Idea	
Pros (Benefits)	**Cons (Limitations)**

VASELINE

THE "WONDER JELLY"

A Case Study Using the Discovery Engineering Process

Lesson Overview

Over 150 years ago, chemist Robert Augustus Chesebrough became aware of a waxlike substance created as a byproduct of oil drilling. He took this raw material and refined it, creating the product that is known today as petroleum jelly or Vaseline. There are many interesting applications for this product. In this lesson, students will explore the insulating properties of petroleum jelly and brainstorm novel product applications for the substance.

Lesson Objectives

By the end of this case study, students will be able to

- Describe how petroleum jelly was discovered.

- Analyze some of the useful properties of petroleum jelly.

- Design a new product that uses petroleum jelly to solve a problem.

The Case Study Approach

This lesson uses a case study approach. Explaining the purpose of case studies will encourage your students to relate to the material and engage with the problem. At the heart of each case study in this book is a true story, one that describes how someone in his or her everyday life or during a routine workday made an observation or did a simple experiment that led to a new insight or discovery. Case studies are designed to get students actively engaged in the process of problem solving. The narrative of the case supplies authentic details that place the student in the role of the inventor and provide scaffolds for critical thinking and deep reflection. A case is more than a paragraph to read or a story to analyze but rather a way of framing problems, synthesizing what is known, and thinking creatively about new applications and solutions. In this lesson, students consider how petroleum jelly was discovered and work together to think about new applications for petroleum jelly to solve real-life problems.

Use of the Case

Due to the nature of these case studies, teachers may elect to use any section of each case for their instructional needs. The sections are sequenced in order (scaffolded) so students think more deeply about the science involved in the case and develop an understanding of engineering in the context of science.

Curriculum Connections

Lesson Integration

You could use this case as a way to integrate engineering into a lesson on chemistry, including discussions of nonpolar hydrophobic compounds and insulating materials.

Related Next Generation Science Standards

PERFORMANCE EXPECTATIONS

- MS-PS1-3. Gather and make sense of information to describe that synthetic materials come from natural resources and impact society.

- MS-PS1-4. Develop a model that predicts and describes changes in particle motion, temperature, and state of a pure substance when thermal energy is added or removed.

- MS-ETS1-1. Define the criteria and constraints of a design problem with sufficient precision to ensure a successful solution, taking into account relevant scientific principles and potential impacts on people and the natural environment that may limit possible solutions.

- HS-ETS1-2. Design a solution to a complex real-world problem by breaking it down into smaller, more manageable problems that can be solved through engineering.

- HS-ETS1-3. Evaluate a solution to a complex real-world problem based on prioritized criteria and trade-offs that account for a range of constraints, including cost, safety, reliability, and aesthetics, as well as possible social, cultural, and environmental impacts.

SCIENCE AND ENGINEERING PRACTICES

- Asking Questions and Defining Problems
- Analyzing and Interpreting Data

- Engaging in Argument From Evidence
- Constructing Explanations and Designing Solutions

CROSSCUTTING CONCEPTS

- Patterns
- Cause and Effect
- Structure and Function

Related National Academy of Engineering Grand Challenges

- Engineer Better Medicines
- Advance Health Informatics
- Engineer the Tools of Scientific Discovery

Lesson Preparation

You will need to make copies of the entire student section for the class. Students will need internet access to read an article and watch a video in the Apply and Analyze section. Alternatively, you can print and distribute copies of the article and project the video for the class. Look at the Teaching Organizer (Table 19.1, p. 320) for suggestions on how to organize the lesson.

Materials

For each group of students:

- Ice water
- Warm water
- Petroleum jelly
- 1000 ml beakers (2 per group)
- 1 graduated cylinder
- 4 zipper sandwich bags
- Plastic spoon
- 2 thermometers
- Scale to measure in grams

- Stopwatch or timing device
- Indirectly vented chemical splash safety goggles (1 pair per student)
- Nonlatex apron (1 per student)
- Nitrile gloves (1 pair per student)

Safety Note for Students: Wear indirectly vented chemical splash safety goggles, a nonlatex apron, and nitrile gloves during the setup, hands-on, and takedown segments of the activity. Immediately wipe up any spilled water on the floor as it is a slip/fall hazard. Use caution when working with glassware, which can shatter if dropped and cut skin. Wash your hands with soap and water immediately after completing this activity.

Time Needed

75 minutes

TABLE 19.1

Teaching Organizer

Section	Time Suggested	Materials Needed	Additional Considerations
The Case	15 minutes	Student packet	Could be read in class or as a homework assignment prior to class
Investigate	20 minutes	Student packet, ice water, warm water, petroleum jelly, 1000 ml beakers (2 per group), 1 graduated cylinder, 4 zipper sandwich bags, plastic spoon, 2 thermometers, scale to measure in grams, stopwatch or timing device, indirectly vented chemical splash safety goggles, nitrile gloves, nonlatex apron	Recommended as an activity students carry out in lab pairs or small groups
Apply and Analyze	10 minutes	Student packet, internet access	Whole-class, small-group, or individual activity
Design Challenge	30 minutes	Student packet	Small-group activity

Teacher Background Information

It may be useful to review the properties of petroleum jelly by watching the short video "The many benefits of petroleum jelly" from the Mayo Clinic (*www.youtube.com/watch?v=ubpsosv7mHM*).

Vocabulary

- hydrophobic
- petroleum

Teacher Answer Key

Recognize, Recall, and Reflect

1. **Why did chemist Robert Chesebrough go to Titusville, Pennsylvania?**

   Chesebrough went to Titusville to determine if any new products could be made from the petroleum recently discovered there.

2. **What did Chesebrough learn from the workers drilling for oil in Titusville?**

   He learned that a waxlike substance would form on the workers' drilling equipment as they drilled. The workers told him that the waxy material helped their cuts and burns heal faster, so they had been applying it to their skin.

3. **Why was it necessary for Chesebrough to work in his lab for several years with the waxy petroleum jelly?**

   Chesebrough spent several years designing a machine that would filter and refine the petroleum jelly in order to create a purified version that he could sell to consumers. Once he had invented his apparatus, he filed a patent and was then ready to manufacture and sell his product.

Questions for Reflection

1. **Look at the data you collected. Did you notice any patterns?**

   Answers will vary, but students should note that the water in the "Petroleum Jelly" set of bags stays warmer for longer.

2. **Did your observations match your prediction? What was different?**

   Answers will vary.

3. **What does the data you collected suggest about the properties of petroleum jelly?**

   Answers will vary, but students might say that petroleum jelly is a good insulator.

4. **Sometimes swimmers who swim in cold water smear petroleum jelly on their bodies. Based on your data, would petroleum jelly help keep them warm?**

   Student data should support the idea that petroleum jelly will help keep swimmers warm in cold water.

5. **What remaining questions do you have about petroleum jelly?**

   Answers will vary.

6. **How would you design an experiment to collect data related to your question?**

   Answers will vary.

Apply and Analyze

1. **Describe what petroleum jelly does when rubbed onto the skin.**

   When applied to the skin, petroleum jelly forms a layer that minimizes heat loss and moisture loss through evaporation. The jelly also forms a thin barrier to the outside environment, which can protect skin from air, moisture, dirt, etc.

2. **What are the benefits of using petroleum jelly for skin care?**

   Petroleum jelly is chemically similar to proteins in our skin and is useful for treating ailments such as chapped lips, dry skin, and rashes like eczema. According to Mayo Clinic dermatologist Dr. Dawn Davis, "It simply sits on top of the skin like a greenhouse roof. So it's like insulating the skin, so that it doesn't lose heat and so it doesn't lose moisture."

3. **Robert Chesebrough called petroleum jelly a wonder product and claimed that it accelerated healing. Based on the information presented by the article and the video, do you think his claim is accurate?**

   Answers will vary. Students may say that petroleum jelly is not a magical curative that heals any wound it is applied to, but it can assist the body's normal healing process when used on specific types of skin ailments. It prevents the area of the skin it is applied to from drying out, which may ease certain types of skin ailments, such as minor burns and rashes. It also forms a protective layer against outside contaminants, which is helpful when wounds are healing.

Reflect

1. **What technologies might need to be developed to create or manufacture this design?**

    Answer depends on the student's design.

2. **What are any constraints or drawbacks you can foresee with implementing this design?**

    Answer depends on the student's design.

3. **Would there be any environmental or human health concerns about this design?**

    Answer depends on the student's design.

Assessment

The Design Challenge can be assessed using the rubric in the appendix (p. 377).

Extensions

This lesson can be followed with lessons about other products that may be created when refining petroleum. It could also be extended by investigating other insulators (such as Styrofoam).

Resources and References

Conger, C. Can petroleum jelly be used as a moisturizer? HowStuffWorks. *https://health. howstuffworks.com/skin-care/moisturizing/products/petroleum-jelly-moisturizer.htm/printable.*

Mayo Clinic. 2017. "Mayo Clinic Minute: The many benefits of petroleum jelly." YouTube video. *www.youtube.com/watch?v=ubpsosv7mHM.*

Panati, C. 2016. *Panati's extraordinary origins of everyday things.* New York: Chartwell Books.

Vaseline. The Vaseline story. *www.vaseline.us/article/vaseline-history.html.*

I'M RUBBER AND YOU'RE GLUE

Vulcanized Rubber

A Case Study Using the Discovery Engineering Process

Introduction

We use rubber in many ways in our everyday lives. The material can be found in everything from the soles of your shoes to food storage containers. Rubber comes from a natural material called latex. However, latex alone is not a very useful material. It tends to melt when it is hot and crack when it is cold. Many people became interested in the potential applications of rubber in the 1800s. But it wasn't until the accidental discovery of a process to stabilize the material that rubber use became widespread.

Lesson Objectives

By the end of this case study, you will be able to

- Describe how vulcanized rubber was made and why it was different from natural rubber.

- Analyze the process of creating synthetic rubber.

- Design your own rubber eraser.

The Case

Read the summary of the article "Charles Goodyear and the Vulcanization of Rubber," published by *connecticuthistory.org*. Then answer the questions that follow.

Latex is the liquid form of sap from a tree found in Brazil. When the sap hardens, it turns into rubber. In the 1800s, many people had staked their fortunes on this new "miracle product." One of those people was Charles Goodyear. Goodyear began his career in rubber after a visit to a rubber company in New York. While he was there, he saw a stack of rubber life preservers. He decided to invent a better inflation tube for these life vests.

When Goodyear returned to the rubber company to show the manager the new tube he had developed, the manager told him that it wasn't the valves that needed to be improved but rather the rubber itself. One story says that the manager took him back to the company warehouse where thousands of rubber items sat melted and stinking. Natural rubber is very sensitive to temperature and, in the early days of rubber manufacturing, products made from natural rubber would melt into blobs in the summer heat. After seeing the ruined rubber in the warehouse, Goodyear developed an obsession with improving the material.

Goodyear spent the next five years trying to develop a stabilizer for rubber. His attempts to improve the durability of the material involved mixing latex with different powders. In order to fund his research, Goodyear sold many of his family's possessions. One tale claims he only refused to sell his wife's china because the teacups could be used to hold the ingredients in his experiments! His passion put his family into debt and landed him in debtors' prison more than once. Goodyear's wife was very supportive of his work and even brought him materials to work with when he was in prison.

Several of Goodyear's experiments seemed to be a success. First, he made shoes using magnesium powder mixed in with the latex. These appeared to be very promising until the weather warmed and all of his shoes melted. His next major "success" came when he tried to use nitric acid to remove paint from one of his rubber concoctions. The rubber turned dry and black. Goodyear threw the piece away but later realized it felt different than the other materials he had created. He used the material to create and sell a bunch of mailbags for the postal service, claiming they were now waterproof. Unfortunately, in warm weather the insides of the bags melted.

After many years of failed attempts, Goodyear made a breakthrough. According to various accounts, the solution came when he accidentally dropped rubber mixed with sulfur onto a hot stove. When he went to scrape it off, he noticed the concoction had hardened but still retained the elasticity that rubber was known for. It took him several months to re-create the process (which became known as vulcanization), but the problem had been solved!

Unfortunately for Goodyear, his fortunes did not improve. There were lawsuits over who owned the patent to his rubber-making process, so he did not see any of the profits. At the age of 59, Goodyear died due to poor health, which had resulted from all the toxic fumes he breathed in during his experiments. His family eventually profited from his work and were able to live the rest of their lives in comfort. After his death, the inventor continued to inspire others and is even the namesake for a famous company that operates today: the Goodyear Tire & Rubber Company.

Recognize, Recall, and Reflect

1. What was the problem with natural rubber products in the 1800s?

2. How did Goodyear attempt to improve natural rubber products?

3. What happened that allowed Goodyear to come up with a solution to improve rubber?

Investigate

In this activity you will make your own rubber erasers.

Materials

For each group of students:

- 10 ml vinegar

- 10 ml water, plus more for dunking

- 10 ml mold-making rubber (e.g., liquid latex rubber like Mold Builder)

- Silicone ice-cube or chocolate mold (or a mold of your own design)

- Disposable coffee stirrer

- Disposable cup for mixing

- Indirectly vented chemical splash safety goggles (1 pair per student)

- Nonlatex apron (1 per student)

- Nitrile gloves (1 pair per student)

- Face shields (1 per student)

Safety Note: If you are using powdered rubber mold, face shields should also be worn. Wear indirectly vented chemical splash safety goggles, a nonlatex apron, and nitrile gloves during the setup, hands-on, and takedown segments of the activity. Follow your teacher's instructions for disposing of waste materials. Wash your hands with soap and water immediately after completing this activity.

Create, Innovate, and Investigate

- Begin by pouring the rubber mold into your disposable cup. What do you notice about the rubber mold? What is the texture? Why do you think that is?

- Next, add the water to your rubber mold and stir. What do you notice? What happens if you stir the mixture too quickly?

- Once your water and rubber are mixed, slowly pour in the vinegar while stirring constantly. How does the consistency change?

- After the vinegar has been added, carefully remove the resulting product from the cup. You will then place it into the silicone mold to shape it. How does the texture feel now?

- Slowly remove your eraser from the silicone mold, being careful not to bend the shape.

- Dunk the eraser into a beaker of water. This will remove any bubbles from your eraser. When the bubbles stop emerging, remove the eraser and set it out to dry (two to three hours).

- Once the eraser has dried, it is ready for use.

Questions for Reflection

1. What did you observe about the changes to the rubber mold as you added water?

2. What did you observe about the changes when you added vinegar?

3. How is the dried eraser different from the original rubber mold?

Apply and Analyze

Rubber is one of the most commonly used products in the world today, found in everything from rubber bands to swim caps to tires. Read the following information about the evolution of the material and explore the web links below. Then answer the questions that follow.

Ancient civilizations in Mexico and Central America were making and using rubber over 3,000 years ago (circa 1800 BC). The word *rubber* is used for a variety of products including natural rubber, latex, and synthetic rubber. When modern manufacturers began making rubber products on a large scale, they found that natural rubber did not react well to hot or cold environments. Charles Goodyear came up with a process called vulcanization, in which rubber was combined with sulfur and heated. The process produced rubber that was much more durable. In the early 1920s, German scientist Hermann Staudinger identified the chemical structure of natural rubber and the properties that gave this rubber its most useful characteristics. Using that information, he developed the first form of synthetic rubber by creating a new polymer. The first large-scale production of his synthetic rubber began in 1939. Now, most synthetic rubber is made from petroleum (oil) products, and it must still go through the vulcanization process when manufactured.

To learn more, check out the links below.

- *https://w3.siemens.com/mcms/sensor-systems/CaseStudies/CS_Butyl_ Rubber_2013-01_en_Web.pdf*

- *www.explainthatstuff.com/rubber.html*

- *www.conserve-energy-future.com/tirerecycling.php*

1. How is natural rubber produced?

2. How does vulcanizing rubber make it stronger?

3. What is the issue with rubber waste? How are manufacturers attempting to solve that problem?

4. What do you see as the best way to solve the issue of rubber waste?

Design Challenge

Engineering is the application of scientific understanding through creativity, imagination, problem solving, and the designing and building of new materials to address and solve problems in the real world. You will be asked to take the science you have learned in this case and design a process or product to address a real-world issue of your choosing.

Engineers use the engineering design process as steps to address a real-world problem (see Figure 20.1, p. 330). You will now use this process as you come up with a new way to use vulcanized rubber. In this case, you are asking the question (Step 1) of how you can design a new use for vulcanized rubber. Drawing on your creativity, you will then brainstorm (Step 2) a new product that uses vulcanized rubber to solve a problem. Afterward, you will create a plan (Step 3) for this new product. Next, you will create a sketch and/or model of your product (Step 4). Then, you will work with your classmates to think about how you would test (Step 5) and refine (Step 6) your product.

1. Ask Questions

Based on your previous research, consider a new problem that may be addressed or product that could be created by using vulcanized rubber. Think about applications where you need a material that can flex, bend, or bounce without melting or cracking. What problem could be solved with vulcanized rubber?

2. Brainstorm and Imagine

Visit the following website, which includes descriptions of different types of synthetic rubber and the ways that they can be used: *https://iisrp.com/synthetic-rubber*. Explore two or three types of synthetic rubber mentioned on the site. Then brainstorm a new application you can create using one of these versions of rubber. (For example, babies often fall down while learning to walk. Perhaps you could use one of the types of rubber to develop a new bumper system that allows babies to bounce back up without being injured in the fall).

FIGURE 20.1

The Engineering Design Process

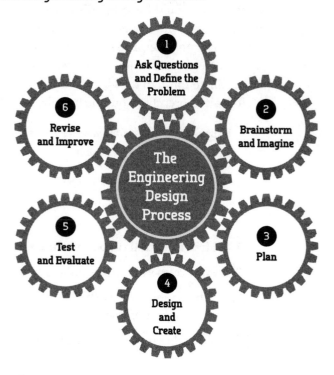

3. Create a Plan

Create a plan for your product. Consider: (1) What is the purpose of the product? (2) What are benefits of the product? (3) What are the limitations of the product? Use the Product Planning Graphic Organizer (p. 332) to help you. (In the Proposed Product Idea section of the organizer, don't forget to include the type of rubber you chose to use and why.)

4. Design and Create

Consider the following questions and considerations for your product and its design.

- How would incorporating vulcanized rubber into your design make the product better?

- Are there any limitations or drawbacks to using vulcanized rubber in your design? If so, how would you overcome them?

- What technologies might need to be developed to create or manufacture this design?

- What are any constraints or drawbacks you can foresee with implementing this design?

- Would there be any safety concerns regarding your product?

Create a sketch of your product design. Make sure your design incorporates your previous research and exploration.

5. Test and Evaluate

Working with your classmates, come up with a way to test your design to see its effectiveness.

6. Revise and Improve

Give your plans to one of your classmates for review. Listen to his or her feedback on your design. What are some ways you can use the input to refine your design? Take some time to revise and make improvements.

Reflect

1. What technologies might need to be developed to create or manufacture this design?

2. What are any constraints or drawbacks you can foresee with implementing this design?

3. Would there be any environmental or human health concerns about the design?

Product Planning Graphic Organizer

Proposed Product Idea	
Pros (Benefits)	**Cons (Limitations)**

I'M RUBBER AND YOU'RE GLUE
VULCANIZED RUBBER

A Case Study Using the Discovery Engineering Process

Lesson Overview

In this lesson, students explore vulcanized rubber. Vulcanized rubber was an accidental discovery that has led to a number of new applications and products including many of the synthetic rubber objects we use in everyday life.

Lesson Objectives

By the end of this case study, students will be able to

- Describe how vulcanized rubber was made and why it was different from natural rubber.

- Analyze the process of creating synthetic rubber.

- Design a rubber eraser.

The Case Study Approach

This lesson uses a case study approach. Explaining the purpose of case studies will encourage your students to relate to the material and engage with the problem. At the heart of each case study in this book is a true story, one that describes how someone in his or her everyday life or during a routine workday made an observation or did a simple experiment that led to a new insight or discovery. Case studies are designed to get students actively engaged in the process of problem solving. The narrative of the case supplies authentic details that place the student in the role of the inventor and provide scaffolds for critical thinking and deep reflection. A case is more than a paragraph to read or a story to analyze but rather a way of framing problems, synthesizing what is known, and thinking creatively about new applications and solutions. In this lesson, students consider how vulcanized rubber was discovered and work together to think about new applications for synthetic rubber to solve real-life problems.

Use of the Case

Due to the nature of these case studies, teachers may elect to use any section of each case for their instructional needs. The sections are sequenced in order (scaffolded) so students think more deeply about the science involved in the case and develop an understanding of engineering in the context of science.

Curriculum Connections

Lesson Integration

You could use this case as a way to integrate engineering into a lesson on physical and chemical changes.

Related Next Generation Science Standards

PERFORMANCE EXPECTATIONS

- MS-PS1-2. Analyze and interpret data on the properties of substances before and after the substances interact to determine if a chemical reaction has occurred.

- MS-PS1-3. Gather and make sense of information to describe that synthetic materials come from natural resources and impact society.

- MS-ETS1-1. Define the criteria and constraints of a design problem with sufficient precision to ensure a successful solution, taking into account relevant scientific principles and potential impacts on people and the natural environment that may limit possible solutions.

- HS-ETS1-2. Design a solution to a complex real-world problem by breaking it down into smaller, more manageable problems that can be solved through engineering.

- HS-ETS1-3. Evaluate a solution to a complex real-world problem based on prioritized criteria and trade-offs that account for a range of constraints, including cost, safety, reliability, and aesthetics, as well as possible social, cultural, and environmental impacts.

SCIENCE AND ENGINEERING PRACTICES

- Analyzing and Interpreting Data
- Engaging in Argument From Evidence
- Constructing Explanations and Designing Solutions

CROSSCUTTING CONCEPT

- Cause and Effect

Related National Academy of Engineering Grand Challenges

- Restore and Improve Urban Infrastructure
- Engineer the Tools of Scientific Discovery

Lesson Preparation

You will need to make copies of the entire student section for the class. Students will need internet access at various points in the lesson. Alternatively, you can print and distribute copies of online content for the class. For the Investigate section, you will need to purchase rubber mold. The rubber mold should be liquid latex rubber like Mold Builder. (Buying the rubber mold in powdered form is not recommended.) Liquid latex rubber can be found in bulk at hardware stores or in craft stores. You may want to premeasure the rubber mold into the disposable cups to reduce waste of the product. If you would like students to design their own molds, you will need to collect supplies such as popsicle sticks, cardboard, tape, or other materials. Students may want to line their own molds with plastic wrap to prevent it from sticking. Petroleum jelly also works well. Look at the Teaching Organizer (Table 20.1, p. 336) for suggestions on how to organize the lesson.

Materials

For each group of students:

- 10 ml vinegar
- 10 ml water, plus more for dunking
- 10 ml mold-making rubber (e.g., liquid latex rubber like Mold Builder)
- Silicone ice-cube or chocolate mold (or a mold of your own design)
- Disposable coffee stirrer
- Disposable cup for mixing
- Indirectly vented chemical splash safety goggles (1 pair per student)
- Nonlatex apron (1 per student)
- Nitrile gloves (1 pair per student)
- Face shields (1 per student)

20

Safety Note for Students: If you are using powdered rubber mold, face shields should also be worn. Wear indirectly vented chemical splash safety goggles, a non-latex apron, and nitrile gloves during the setup, hands-on, and takedown segments of the activity. Follow your teacher's instructions for disposing of waste materials. Wash your hands with soap and water immediately after completing this activity.

Time Needed

3 class periods

TABLE 20.1

Teaching Organizer

Section	Time Suggested	Materials Needed	Additional Considerations
The Case	5 minutes	Student packet	Could be read in class or as a homework assignment prior to class
Investigate	2 class periods	Student packet, 10 ml vinegar, 10 ml water plus more for dunking, 10 ml mold-making rubber, silicone ice-cube or chocolate mold (or mold of your own design), disposable coffee stirrer, disposable cup for mixing, indirectly vented chemical splash safety goggles, nitrile gloves, nonlatex apron, face shields	This can be done in groups with more than one student filling the same ice-cube mold. The activity may take more time if students design their own mold.
Apply and Analyze	10 minutes	Student packet, internet access	Small-group or individual activity
Design Challenge	45 minutes	Student packet, internet access	Small-group activity

Teacher Background Information

There are a number of resources and videos about vulcanized rubber available on the internet. You may want to observe the creation of vulcanized rubber on sites such as YouTube prior to using the case. Students often have questions about chemical and physical change. It may be useful to go over these resources from the National Science Teachers Association as a review:

- Book chapter on understanding chemical changes
  http://common.nsta.org/resource/?id=10.2505/9781935155232.12

- Journal article on chemical and physical changes
  http://common.nsta.org/resource/?id=10.2505/4/ss09_033_02_54

- Book on rubber versus glass
  www.nsta.org/publications/press/extras/rubber.aspx

Vocabulary

- latex
- polymer

- stabilizer
- vulcanized

Teacher Answer Key

Recognize, Recall, and Reflect

1. **What was the problem with natural rubber products in the 1800s?**

    They melted in hot weather and cracked in cold weather.

2. **How did Goodyear attempt to improve natural rubber products?**

    He mixed different powders into the latex in order to strengthen it.

3. **What happened that allowed Goodyear to come up with a solution to improve rubber?**

    He accidentally dropped latex mixed with sulfur on a hot stove where it turned into vulcanized rubber.

Questions for Reflection

1. **What did you observe about the changes to the rubber mold as you added water?**

    The rubber mold became a thinner liquid similar to milk.

2. **What did you observe about the changes when you added vinegar?**

    The rubber became solid but left excess liquid in the cup.

3. **How is the dried eraser different from the original rubber mold?**

    It is a flexible, solid rubber rather than a liquid that resembles glue.

Apply and Analyze

1. **How is natural rubber produced?**

   Natural rubber is made by collecting sap from a tree. Then it can be rolled into many shapes. However, natural rubber is very weak, so it is often vulcanized to make it stronger.

2. **How does vulcanizing rubber make it stronger?**

   In natural rubber, latex molecules are tangled and weakly held together. Adding sulfur bridges the molecules and holds them together more strongly.

3. **What is the issue with rubber waste? How are manufacturers attempting to solve that problem?**

   There is a lot of waste produced, and it is toxic. It is often burned, which releases toxins into the air. Manufacturers are attempting to find new ways to recycle rubber.

4. **What do you see as the best way to solve the issue of rubber waste?**

   Answers may vary.

Reflect

1. **What technologies might need to be developed to create or manufacture this design?**

   Student answers will vary.

2. **What are any constraints or drawbacks you can foresee with implementing this design?**

   Student answers will vary.

3. **Would there be any environmental or human health concerns about the design?**

   Student answers will vary.

Assessment

The Design Challenge can be assessed using the rubric in the appendix (p. 377).

Extensions

This lesson can be followed with the case study "That's a Wrap: Plastic" in Chapter 9 (p. 139).

Resources and References

Goodyear. The Charles Goodyear story. *https://corporate.goodyear.com/en-US/about/history/charles-goodyear-story.html* (accessed January 5, 2017).

Grady, E. How to make rubber erasers. eHow. *www.ehow.com/how_8256496_make-rubber-erasers.html* (accessed January 5, 2017).

Haysom, J., and M. Bowen. 2010. Understanding chemical changes. In *Predict, observe, explain: Activities enhancing scientific understanding*, 241–260. Arlington, VA: NSTA Press.

International Institute of Synthetic Rubber Producers. Synthetic rubber. *https://iisrp.com/synthetic-rubber* (accessed January 5, 2017).

Kukreja, R. Tire recycling. Conserve Energy Future. *www.conserve-energy-future.com/tirerecycling.php* (accessed January 5, 2017).

Lowery, L. 2014. *Rubber vs. glass*. Arlington, VA: NSTA Press.

McIntosh, J., S. White, and R. Suter. 2009. Science sampler: Enhancing student understanding of physical and chemical changes. *Science Scope* 33 (2): 54–58.

Siemens. Production of synthetic rubber. *https://w3.siemens.com/mcms/sensor-systems/CaseStudies/CS_Butyl_Rubber_2013-01_en_Web.pdf* (accessed January 5, 2017).

Somma, A. M. Charles Goodyear and the vulcanization of rubber. Connecticut Humanities. *https://connecticuthistory.org/charles-goodyear-and-the-vulcanization-of-rubber* (accessed January 5, 2017).

Woodford, C. 2017. Rubber. Explain That Stuff. *www.explainthatstuff.com/rubber.html*.

PEERING INTO THE UNKNOWN

The Discovery of X-Rays

A Case Study Using the Discovery Engineering Process

Introduction

What are invisible to the naked eye and can be used to look inside structures like the human body and apartment buildings without leaving a trace? This isn't something from science fiction—we're talking about x-rays. Since their discovery in 1895, x-rays have captured our imagination and led to many extraordinary innovations. Scientists have taken advantage of the unique properties of x-rays to increase our understanding of the world, using x-ray technology to make advancements in everything from high-energy astrophysics to atomic modeling.

X-rays are a type of electromagnetic radiation that have more energy than radio waves, microwaves, infrared radiation, visible light, and ultraviolet light. Though x-rays might seem mysterious, they are used in many common applications that impact you on a daily basis.

Lesson Objectives

By the end of this case study, you will be able to

- Describe what x-rays are.

- Demonstrate how medical professionals use x-rays to create medical images of the human body.

- Design a new application for an x-ray technology that solves a problem.

The Case

This case summarizes how x-rays were accidentally discovered at the end of the 19th century and shares early reactions to the new, unexpected technology. Once you've read the account, answer the questions that follow.

Wilhelm Conrad Röntgen (also spelled "Roentgen") was a German physicist who was awarded the first Nobel Prize in Physics in 1901. Eighty years after his death, he was honored again by having an element on the periodic table named for him. Despite these extraordinary achievements, he had a rather surprising start to his career. While attending high school in the Netherlands, Röntgen was blamed for a prank committed by another student and expelled. Not having a high school diploma caused him some difficulty early on, but he found a way to attend university, earn his doctorate, and become a professor of physics.

While working as a professor at the University of Würzburg, Röntgen conducted many experiments with electricity and light. He used a special device in his experiments called a Crookes tube, which is an electrical discharge tube that is mostly empty of air and features a cathode (negative electrode) and an anode (positive electrode) at either end. At the time, it was understood that cathode rays (or streams of what would become known as electrons) traveled in a straight line from the device's cathode when a high-voltage electrical current was applied between the anode and cathode. When the device was turned on, the cathode rays would strike the wall across from the cathode, and the glass of the tube would fluoresce, glowing green. (Figure 21.1 shows photos of a Crookes tube turned off [left] and in use [right].) In the mid-1890s, scientists like Röntgen were still figuring out what was going on inside the tube to make this happen. (See the What Caused the Crookes Tube to Glow Green? sidebar [p. 344] for more information.)

FIGURE 21.1

A Crookes Tube Turned Off and a Crookes Tube in Use

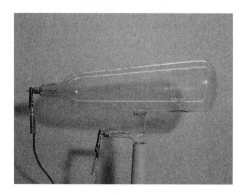 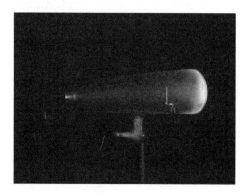

For one of his experiments in 1895, Röntgen completely covered the Crookes tube in heavy black cardboard to block out the green glow of the tube. He then turned off the lights in his lab. Röntgen observed that a screen in the lab that was coated in barium platinocyanide (a salt with fluorescent properties) glowed faintly if placed in the path of the cathode rays emanating from the tube. The screen would even glow if placed around nine feet away from the cathode. Röntgen found this very surprising: Based on his understanding of cathode rays, they could not be causing the screen to glow from several feet away. What else could be making the screen glow at such a distance? What wasn't he seeing?

FIGURE 21.2

Röntgen's Lab at the University of Würzburg

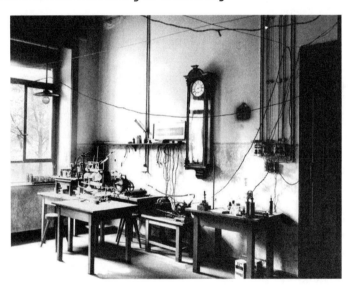

Röntgen spent the next several weeks conducting carefully planned experiments in his lab (Figure 21.2) to come up with a convincing explanation for the glowing screen. He concluded that he had discovered a new kind of invisible ray that was created when the Crookes tube was turned on. He called this discovery x-radiation or x-rays, using the mathematical variable x, which indicates when something is unknown.

Röntgen learned from his experiments that x-rays could travel long distances and pass through some solid materials but not others. He tested many different types of materials to see which would block the x-rays. His most impactful discovery came when he placed his hand in front of the screen and saw the shadows cast by his bones. Röntgen realized that the rays could more easily pass through flesh than bone, allowing him to see *inside* his own body! Using a photographic plate,

FIGURE 21.3

Röntgen's X-Ray Image of Wife's Hand

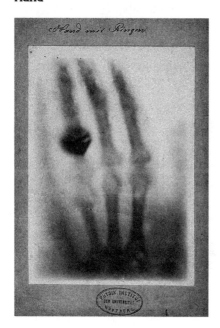

Röntgen took an image of his wife Bertha's hand (Figure 21.3), which captured the dark shadows of her bones and her wedding ring surrounded by the lighter shadows cast by the flesh of her hand. The field of x-ray imaging had begun!

What Caused the Crookes Tube to Glow Green?

The cause of the Crookes tube's greenish glow is now understood. In 1897, British scientist J. J. Thomson tried a different experiment with the Crookes tube and learned that cathode rays behave like negatively charged particles. He called these newly found particles electrons. (Thomson received the Nobel Prize in Physics in 1906 for discovering electrons.) Fast-moving electrons leaving the cathode run into the glass wall of the Crookes tube and transfer energy to the atoms in the glass on impact. The atoms' electrons absorb energy and jump to a higher energy level. When these electrons release their extra energy and return to a lower energy state, they emit a greenish light. X-rays are also created in this process. The x-rays are generated when fast-moving electrons quickly change their speed (for example, when they smash into the glass wall of the tube).

Recognize, Recall, and Reflect

1. What difficulty did Röntgen have to overcome early in his life?

2. How are x-rays produced by the Crookes tube?

3. What happened in Röntgen's lab to reveal that x-rays could be used to see inside the human body?

Dive Deeper: Early Experimentation With X-Ray Imaging

News of Röntgen's discovery quickly spread, sparking the imagination of scientists and the general public. Röntgen believed that his discovery belonged to the world and refused to patent it, which allowed other scientists and entrepreneurs to quickly improve on his x-ray imaging technology.

The technology began showing up in daily life. Photography studios opened to take customers' "bone portraits"; some shoe stores had portable x-ray machines installed so customers could look at the bones in their feet while trying on shoes; and rumors spread that x-ray glasses could allow you to see through walls! X-ray machines were immediately used for medical applications; early cases included diagnosing bone fractures and locating metal objects such as bullets.

At the time it was not well understood that x-rays have the potential to harm living tissue and exposure should be limited. Widespread experimentation with the new technology resulted in many early stories of radiation burns and hair loss, and extended exposure over many years led to deaths. The early x-ray devices exposed users to a very high dose of x-rays, often for many minutes (or hours) at a time. An x-ray device from 1896 used a radiation dose 10 times higher than modern x-ray equipment and required an exposure time of tens of minutes. The resulting skin dose was about 100 times greater for these early machines than what it is today. In modern times, x-ray exposure is quite low during a medical procedure, with patients being exposed to radiation for merely tens of milliseconds. And with such low doses of radiation, the benefits of an exposure are now much greater than the risks involved.

Investigate

X-ray imaging has revolutionized the diagnosis and treatment of many medical conditions by allowing medical professionals to look inside the human body and determine the location of a broken bone, foreign object, or evidence of disease. In this activity, you will explore what considerations are important when creating a clear x-ray image. Then you will design your own experiment with a variable you would like to test. To complete this activity, it is important to have some background information about how medical x-ray imaging, called radiography, works.

To create a radiograph (an x-ray image), a patient is placed between an x-ray source and an x-ray detector, and the x-rays pass through the body as they travel to the detector. Some body tissues absorb more x-rays than others, depending on their radiological density. The density of the tissue and the atomic number of the atoms in the tissue both determine the amount of x-rays the tissue will absorb. X-ray photons carry a lot of energy and an atom with a higher atomic number is more likely to absorb the x-rays, because larger atoms have greater energy differences between orbitals. For example, bones absorb more x-rays than other bodily tissues because they contain a lot of calcium. Calcium has an atomic number higher than atoms found in other types of human tissue.

The following materials are listed from least radiologically dense to most dense. These objects absorb different amounts of x-rays and look different from one another on a radiograph.

1. Air (found in the lungs, the stomach, the intestine, etc.)

2. Fat

3. Soft tissue (e.g., heart, kidney, muscles)

4. Bones

5. Dense foreign bodies (e.g., metal plates and implants)

Using what you have learned, you will now gather the following materials and carry out the activity.

Materials

For each group of students:

- Ring stand with clamp

- At least 1 light source (e.g., lightbulb, flashlight)

- Several pieces of white paper

- Pencils

- 1 set of shapes printed on transparency film and cut out

- Meterstick

- Safety glasses or goggles (1 pair per student)

Safety Note: Incandescent light bulbs can become hot if turned on for long periods of time. If you are using an incandescent bulb as your light source, be mindful and do not touch the bulb. Wear safety goggles or glasses during the setup, hands-on, and takedown segments of the activity. Use only GFI-protected circuits when using electrical equipment, and keep away from water sources to prevent shock. Wash your hands with soap and water immediately after completing this activity.

Create, Innovate, and Investigate

- Begin by examining the different cut-out shapes of transparency film. Arrange the shapes in order from most opaque (blocks the most light) to least opaque. Use the key on page 353 to help refer to the opacity levels (labeled A through E) of the different shapes. What do you notice when you overlap the shapes?

- Set up your experiment:

  o Place a sheet of white paper on the floor. This is your "screen" where your "x-ray image" will appear.

○ Use the ring stand and clamp to mount your light source. It should be pointed down at your paper and located about a meter off the floor. The lightbulb is your "x-ray source." Measure and record the distance from the bulb to the paper.

○ The shapes cut out of the transparency film represent different tissues found in the human body. The large circles are a background tissue, like fat. The squares of different sizes and opacities represent other structures that may be found in the body (e.g., bones and organs). Make a "body model" by arranging several of the squares onto one of the circles. (See Figure 21.4 for an example.)

FIGURE 21.4

Transparency Film Shapes

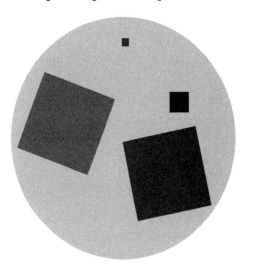

- When your class is ready, your teacher will lower the classroom lights and close the window blinds. Then you will turn on your light source. Hold your group's body model flat with one hand on each side of the circle and place it in between the light and the paper so your model casts shadows onto the paper. (Avoid blocking the lights with your hands as you hold the model.)

- Move your body model in a straight line between the light source and the paper. What do you observe happening to the x-ray image as the model moves up and down?

- Pick three different heights to hold the body model still while you trace the model's shadows onto your paper. Record any observations you make about the dark and light areas of shadow, and the quality of the image of the circle and squares. For each tracing, make sure you record (1) the distance between the body model and the paper and (2) the distance between the light source and the paper.

- Create a new body model and repeat the experiment. Try using shapes of different opacities and sizes. What happens to the shadows if you overlap several shapes? What might this represent in the human body?

- If you have time and the supplies available, try this experiment with a light source of a different diameter. You might exchange a small flashlight with an incandescent light bulb, for example.

Questions for Reflection

1. How did changing the distance between the body model and the paper change the shadows you traced? Describe what you observed.

2. Were the shadows of some squares more easily distinguished from the background circle? How did using squares of different sizes and opacities change how easily you were able to observe them?

3. What would happen to the shadows if you tilted the body model and held it at an angle?

4. Using this experiment as a model for x-ray imaging, what factors should a medical professional consider when creating a radiograph of a three-dimensional body?

5. Design a new experiment using these (or similar) materials. What variables would you test and why?

Apply and Analyze

Crystals and X-Rays

Röntgen's discovery of x-rays in 1895 was major news. Medical professionals immediately started to use x-rays in the treatment of patients, but physicists still could not explain what the rays were. Many scientists worked on this important problem. In 1912, German physicist Max von Laue put together two ideas and created a new experiment to test his theory about x-rays. He knew that other scientists had hypothesized that x-rays might be a form of electromagnetic radiation with a wavelength shorter than visible light. He also knew that when waves interact with empty spaces (slits) that are similar in size to their wavelength, the waves get deflected in specific ways and create unique patterns. In 1912, it was believed (but not yet proven) that crystals were made from regular patterns of repeating atoms, which meant there would be empty spaces between the atoms. Von Laue put this all together and realized that if x-rays were waves and if crystals had regular structures, then a crystal might have just the right amount of space between its atoms to interact with x-rays and create diffraction patterns. Von Laue tried out his experiment by shining x-rays onto a crystal of copper sulphate and was excited to see a pattern of bright spots emerge on a photographic plate! With just one experiment, von Laue had shown that x-rays were a form of electromagnetic radiation *and* he had determined how atoms are arranged in crystals. In 1914, he was awarded the Nobel Prize in Physics for this work.

The idea that x-rays could be used to determine the arrangement of atoms in crystal structures was groundbreaking. It inspired many other scientists to do their own investigations into the subject matter, leading to even more discoveries related to crystallography (or the study of crystals).

Watch the video "Celebrating Crystallography - An animated adventure" by the Royal Institution in London (*www.youtube.com/watch?v=uqQlwYv8VQI*) to find out about the early work of x-ray crystallographers. Then answer the following questions.

1. How many Nobel Prizes have been awarded for work related to x-ray crystallography?

2. What did William and Lawrence Bragg contribute to the field of x-ray crystallography?

3. Why is it useful to understand the structure of materials?

Rosalind Franklin

Watch the video "Rosalind Franklin: DNA's Unsung Hero" by TED-Ed (*https://ed.ted.com/lessons/rosalind-franklin-dna-s-unsung-hero-claudio-l-guerra*) to learn about Rosalind Franklin, a British chemist with expertise in x-ray crystallography. Once you've finished the video, answer the questions below.

1. How were x-rays involved in the discovery of the structure of DNA?

2. What role did scientists James Watson and Francis Crick play in the discovery of the structure of DNA?

3. What barriers did Rosalind Franklin face as a woman who wanted to study science?

Design Challenge

Engineering is the application of scientific understanding through creativity, imagination, problem solving, and the designing and building of new materials to address and solve problems in the real world. You will be asked to take the science you have learned in this case and design a process or product to address a real-world issue of your choosing.

Engineers use the engineering design process as steps to address a real-world problem (see Figure 21.5, p. 350). You will now use this process as you come up with a new way to use x-rays. In this case, you are asking the question (Step 1) of how you can design a new use for x-rays. Drawing on your creativity, you will then brainstorm (Step 2) a new product that uses x-rays to solve a problem. Afterward, you will create a plan (Step 3) for this new product. Next, you will create a sketch and/or model of your product (Step 4). Then, you will work with your classmates to think about how you would test (Step 5) and refine (Step 6) your product.

1. Ask Questions

Based on your previous research, consider a new problem that may be addressed or product that could be created by using x-rays. Think about applications where you would need to see inside or through objects, or where you would need to detect objects that emit high-energy radiation. What problem could be solved with x-rays?

2. Brainstorm and Imagine

Brainstorm a specific new application for x-rays. (For example, people who play in ball pits may lose their keys or phones when they are jumping around. A special x-ray camera could help find these lost objects.) What products may be useful using this technology?

FIGURE 21.5

The Engineering Design Process

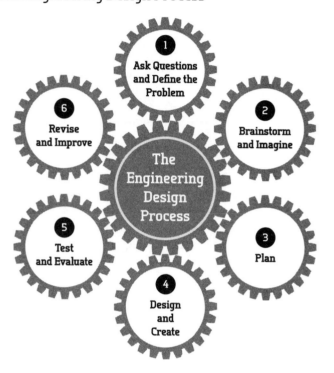

3. Create a Plan

Create a plan for a your product. Consider: (1) What is the purpose of the product? (2) What are benefits of the product? (3) What are the limitations of the product? Use the Product Planning Graphic Organizer (p. 352) to help you.

4. Design and Create

Consider the following questions and considerations for your product and its design.

- How would incorporating x-rays into your design make the product better?

- Are there any limitations or drawbacks to using x-rays in your design? If so, how would you overcome them?

- What technologies might need to be developed to create or manufacture this design?

- What are any constraints or drawbacks you can foresee with implementing this design?

- Would there be any safety concerns regarding your product?

Create a sketch of your product design. Make sure your design incorporates your previous research and exploration.

5. Test and Evaluate

Working with your classmates, come up with a way to test your design to see its effectiveness.

6. Revise and Improve

Give your plans to one of your classmates for review. Listen to his or her feedback on your design. What are some ways you can use the input to refine your design? Take some time to revise and make improvements.

Reflect

1. What technologies might need to be developed to create or manufacture this design?

2. What are any constraints or drawbacks you can foresee with implementing this design?

3. Would there be any environmental or human health concerns about using x-rays in this way?

Product Planning Graphic Organizer

Proposed Product Idea	
Pros (Benefits)	**Cons (Limitations)**

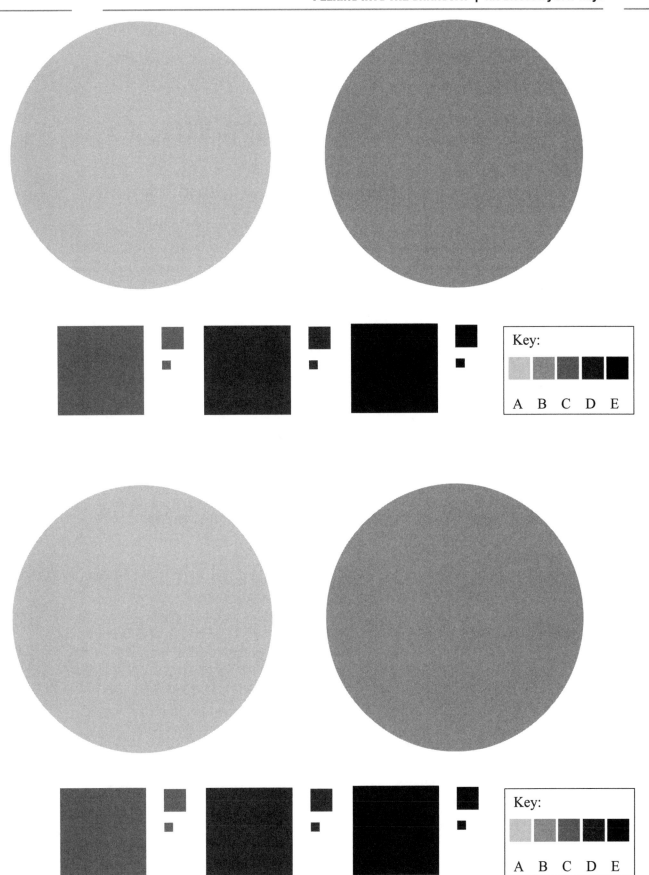

Key:

A	B	C	D	E

PEERING INTO THE UNKNOWN
THE DISCOVERY OF X-RAYS

A Case Study Using the Discovery Engineering Process

Lesson Overview

In this lesson, students explore what x-rays are and how they were discovered. They also learn about various uses for x-rays (e.g., medical imaging, x-ray crystallography) and explore what considerations are important when creating a clear x-ray image. Finally, students design a new application for x-ray technology.

Lesson Objectives

By the end of this case study, students will be able to

- Describe what x-rays are.

- Demonstrate how medical professionals use x-rays to create medical images of the human body.

- Design a new application for an x-ray technology that solves a problem.

The Case Study Approach

This lesson uses a case study approach. Explaining the purpose of case studies will encourage your students to relate to the material and engage with the problem. At the heart of each case study in this book is a true story, one that describes how someone in his or her everyday life or during a routine workday made an observation or did a simple experiment that led to a new insight or discovery. Case studies are designed to get students actively engaged in the process of problem solving. The narrative of the case supplies authentic details that place the student in the role of the inventor and provide scaffolds for critical thinking and deep reflection. A case is more than a paragraph to read or a story to analyze but rather a way of framing problems, synthesizing what is known, and thinking creatively about new applications and solutions. In this lesson, students consider how x-rays were discovered and work together to think about new applications for x-rays to solve real-life problems.

Use of the Case

Due to the nature of these case studies, teachers may elect to use any section of each case for their instructional needs. The sections are sequenced in order (scaffolded), so students think more deeply about the science involved in the case and develop an understanding of engineering in the context of science.

Curriculum Connections

Lesson Integration

You could use this case as a way to integrate engineering into a lesson on electromagnetic waves, atomic structure, or anatomy.

Related Next Generation Science Standards
PERFORMANCE EXPECTATIONS

- MS-PS1-1. Develop models to describe the atomic composition of simple molecules and extended structures.

- MS-PS4-1. Use mathematical representations to describe a simple model for waves that includes how the amplitude of a wave is related to the energy in a wave.

- MS-PS4-2. Develop and use a model to describe that waves are reflected, absorbed, or transmitted through various materials.

- MS-ETS1-1. Define the criteria and constraints of a design problem with sufficient precision to ensure a successful solution, taking into account relevant scientific principles and potential impacts on people and the natural environment that may limit possible solutions.

- HS-PS2-6. Communicate scientific and technical information about why the molecular-level structure is important in the functioning of designed materials.

- HS-PS4-3. Evaluate the claims, evidence, and reasoning behind the idea that electromagnetic radiation can be described either by a wave model or a particle model, and that for some situations one model is more useful than the other.

- HS-PS4-4. Evaluate the validity and reliability of claims in published materials of the effects that different frequencies of electromagnetic radiation have when absorbed by matter.

- HS-ETS1-2. Design a solution to a complex real-world problem by breaking it down into smaller, more manageable problems that can be solved through engineering.

SCIENCE AND ENGINEERING PRACTICES

- Asking Questions and Defining Problems
- Developing and Using Models
- Planning and Carrying Out Investigations
- Analyzing and Interpreting Data
- Engaging in Argument From Evidence
- Constructing Explanations and Designing Solutions

CROSSCUTTING CONCEPTS

- Cause and Effect
- Scale, Proportion, and Quantity
- Structure and Function

Related National Academy of Engineering Grand Challenges

- Engineer Better Medicines
- Advance Health Informatics
- Engineer the Tools of Scientific Discovery

Lesson Preparation

You will need to make copies of the entire student section for the class. For the Investigate section, you will need to print out the circles and squares of various sizes and opacities (p. 353) on transparency sheets. Each group of students should receive one set of cut-out shapes to work with. (There are two sets per 8.5- × 11-inch sheet of transparency.) Students will need internet access to watch videos for the Apply and Analyze section. Alternatively, you can project the videos for the class. Look at the Teaching Organizer (Table 21.1) for suggestions on how to organize the lesson.

Materials

For each group of students:

- Ring stand with clamp

- At least 1 light source (e.g., lightbulb, flashlight)
- Several pieces of white paper
- Pencils
- 1 set of shapes printed on transparency film and cut out (see p. 353)
- Meterstick
- Safety glasses or goggles (1 pair per student)

Safety Note: Incandescent light bulbs can become hot if turned on for long periods of time. If you are using an incandescent bulb as your light source, be mindful and do not touch the bulb. Wear safety goggles or glasses during the setup, hands-on, and takedown segments of the activity. Use only GFI-protected circuits when using electrical equipment, and keep away from water sources to prevent shock. Wash your hands with soap and water immediately after completing this activity.

Time Needed

100 minutes

TABLE 21.1

Teaching Organizer

Section	Time Suggested	Materials Needed	Additional Considerations
The Case	15 minutes	Student packet	Could be read in class or as a homework assignment prior to class
Dive Deeper	5 minutes	Student packet	Could be read in class or as a homework assignment prior to class
Investigate	30 minutes	Student packet, ring stand with clamp, at least 1 light source (e.g., lightbulb, flashlight), several pieces of white paper, pencils, 1 set of shapes printed on transparency film and cut out, meterstick, safety glasses or goggles	Small-group activity
Apply and Analyze	20 minutes	Student packet, internet access	Whole-class, small-group, or individual activity
Design Challenge	30 minutes	Student packet	Small-group activity

Vocabulary

- anode
- atom
- atomic number
- cathode
- current
- diffract
- diffraction pattern
- electrode
- electromagnetic radiation
- electron
- energy level
- fluorescent/fluoresce
- opaque/opacity
- orbital
- radiograph
- radiography
- radiological density
- x-ray

Teacher Answer Key

Recognize, Recall, and Reflect

1. **What difficulty did Röntgen have to overcome early in his life?**

   Röntgen was unfairly expelled from his high school, which made it harder to get into university.

2. **How are x-rays produced by the Crookes tube?**

   X-rays are produced when electrons that are moving very fast are suddenly slowed down or stopped. In the Crookes tube this happens when the electrons are brought to rest by slamming into the glass wall.

3. **What happened in Röntgen's lab to reveal that x-rays could be used to see inside the human body?**

   When Röntgen placed his hand between the Crookes tube and the fluorescent screen, he allowed the x-rays to pass through his hand and create a shadow on the screen. He noticed his bones cast a darker shadow than his flesh, which gave him a view of the inside of his body.

Questions for Reflection

1. **How did changing the distance between the body model and the paper change the shadows you traced? Describe what you observed.**

   The farther the body model was from the screen, the larger (more magnified) but less sharp the image appeared. The closer the body model was to the screen, the smaller and sharper the image became. The shadows of the different shapes are easier to tell apart when their image is sharp and not blurry.

2. **Were the shadows of some squares more easily distinguished from the background circle? How did using squares of different sizes and opacities change how easily you were able to observe them?**

   It is more difficult to identify the boundaries between shapes that are small in size and of similar opacities, especially as the body model moves away from the screen. Larger squares with more contrast between the shape and the background circle are easiest to identify.

3. **What would happen to the shadows if you tilted the body model and held it at an angle?**

   The shadows would appear to change shape.

4. **Using this experiment as a model for x-ray imaging, what factors should a medical professional consider when creating a radiograph of a three-dimensional body?**

   If a medical professional is trying to image a feature in a body that is located behind another tissue, they may need to position the patient in several different ways and take an image from several angles to better see what they are trying to identify. To get the clearest image, they should place their patient as close to the screen as possible.

5. **Design a new experiment using these (or similar) materials. What variables would you test and why?**

   Responses will vary.

Apply and Analyze
CRYSTALS AND X-RAYS

1. **How many Nobel Prizes have been awarded for work related to x-ray crystallography?**

   Twenty-eight Nobel Prizes have been awarded for work related to x-ray crystallography.

2. **What did William and Lawrence Bragg contribute to the field of x-ray crystallography?**

   William and Lawrence Bragg developed an equation, Bragg's law, which made it possible to determine how the spacing between the spots in the diffraction pattern relate to the arrangement of the atoms in the crystal.

3. **Why is it useful to understand the structure of materials?**

Understanding a material's structure can help scientists understand why certain molecules behave the way they do.

ROSALIND FRANKLIN

1. **How were x-rays involved in the discovery of the structure of DNA?**

While working at King's College, Rosalind Franklin projected x-rays onto crystals of DNA to create images of the diffraction pattern caused by the x-rays interacting with the crystals. She then used her resulting data to calculate the structure of the DNA molecule, determining that it was a double helix.

2. **What role did scientists James Watson and Francis Crick play in the discovery of the structure of DNA?**

Watson and Crick used Franklin's data (an x-ray crystallography image known as Photo 51) without her knowledge to figure out the structure of DNA. They were awarded the Nobel Prize in Physiology or Medicine in 1962 for their work on DNA.

3. **What barriers did Rosalind Franklin face as a woman who wanted to study science?**

It was uncommon for women to have careers in science during the time period in which Rosalind Franklin lived. Franklin was isolated from her colleagues and treated unfairly. One of her colleagues even showed some of her data to other scientists (Watson and Crick), who used the data without asking permission and without giving Franklin credit for her work.

Reflect

1. **What technologies might need to be developed to create or manufacture this design?**

Answers will vary depending on the students' designs.

2. **What are any constraints or drawbacks you can foresee with implementing this design?**

Answers will vary depending on the students' designs.

3. **Would there be any environmental or human health concerns about the design?**

Answers will vary depending on the students' designs.

Assessment

The Design Challenge can be assessed using the rubric in the appendix (p. 377).

Extensions

This lesson can be followed with lessons about electromagnetic radiation, other discoveries made with x-rays, subatomic particles and atomic structures, a more detailed investigation of how x-rays are used in the medical professions, and astronomical sources of x-rays.

Resources and References

Beiser, A. 2003. *Concepts of Modern Physics.* 6th ed. New Delhi: Tata McGraw-Hill Publishing Company Limited.

Bertsch, G., J. Trefil, and S. Bertsch McGrayne. 2018. "Atom: Matter." In *Encyclopaedia Britannica. www.britannica.com/science/atom/Discovery-of-electrons.*

Chodos, A., ed. 2001. November 8, 1895: Roentgen's discovery of x-rays. *APS News* 10 (10): 2. *www.aps.org/publications/apsnews/200111/upload/nov01.pdf.*

The Columbia Encyclopedia, 6th ed. "Crookes tube." *www.encyclopedia.com/reference/ encyclopedias-almanacs-transcripts-and-maps/crookes-tube* (accessed January 12, 2018).

Guerra, C. 2016. "Rosalind Franklin: DNA's unsung hero." TED-Ed video. *https://ed.ted.com/ lessons/rosalind-franklin-dna-s-unsung-hero-claudio-l-guerra.*

Kemerink, M., T. J. Dierichs, J. Dierichs, H. J. M. Huynen, et al. 2011. Characteristics of a first-generation x-ray system. *Radiology* 259 (2): 534–539.

National Institute of Biomedical Imaging and Bioengineering. 2017. X-rays. NIH. *www. nibib.nih.gov/science-education/science-topics/x-rays.*

Nobel Media AB. 2014. Max von Laue biographical. *www.nobelprize.org/prizes/physics/1914/ laue/biographical.*

Nobel Media AB. 2014. The Nobel Prize in Physics 1901. *www.nobelprize.org/prizes/ physics/1901/summary.*

Nobel Media AB. 2014. Wilhelm Conrad Röntgen biographical. *www.nobelprize.org/prizes/ physics/1901/rontgen/biographical.*

Nobel Media AB. 2014. X-ray's identity becomes crystal clear. *www.nobelprize.org/prizes/ physics/1914/perspectives.*

O'Rahilly, R., F. Müller, S. Carpenter, and R. Swenson. 1982. *Basic human anatomy: A regional study of human structure.* Philadelphia: W.B. Saunders.

The Royal Institution. 2013. "Celebrating Crystallography - An animated adventure." YouTube video. *www.youtube.com/watch?v=uqQlwYv8VQI.*

Spiegel, P. K. 1995. The first clinical x-ray made in America—100 years. *American Journal of Roentgenology* 164 (1): 241–243.

22

VELCRO

Engineering Mimics Nature

A Case Study Using the Discovery
Engineering Process

Introduction

Imagine a world without zippers, buttons, or Velcro to fasten your clothing. Throughout history, people have designed new ways to keep clothing closed and fitting comfortably. The discovery of Velcro changed the way clothes, shoes, and even suitcases and purses open and close. Velcro is a unique type of hook and loop tape that was inspired by cockleburs in nature. (Cockleburs [Figure 22.1] are seeds covered by hard, curved spines that are difficult to remove if they become hooked onto clothing.) This type of engineering in which objects in nature are used as models for new products is known as biomimicry.

FIGURE 22.1

Cocklebur

Lesson Objectives

By the end of this case study, you will be able to

- Describe the hook-and-loop system that makes up Velcro.

- Analyze how hook-and-loop tape functions to keep materials together and how it is able to repeatedly open and close.

- Design a new application for hook-and-loop tape.

The Case

Read the following description of how Velcro was discovered.

One day in the 1940s, George de Mestral was hunting in the Jura Mountains of Switzerland when he observed that his pants were coated with little cockleburs. He checked his dog and noticed that the dog's fur was covered with cockleburs, too. De Mestral was intrigued by the idea that these little seed pods could stick to his pants so effectively. When he returned to work, he looked at one of the cockleburs under a microscope and saw that tiny hooks on the seed pod were looped into the threads on his pants.

Typically about an inch long, the most common type of cocklebur is from the plant *Xanthium strumarium.* Just one plant can produce many of these spine-covered burs. A single bur contains two seeds that are carried to different areas by animals that happen to bump up against the plant. The cocklebur sticks to the animal's fur until it is dislodged at some point during the animal's wanderings. This mechanism allows for the scattering of cocklebur seeds. And due to the way in which its seeds get dispersed, the cocklebur has been nicknamed "nature's hitchhiker."

Once de Mestral saw the cocklebur hook under his microscope, he decided to make a new type of fastener that mimicked the hooks on the bur and the fibers of his pants. He called this new hook-and-loop fastener Velcro, which is a combination of the words *velvet* and *crochet.* De Mestral went on to obtain a patent in Switzerland for his new fastener, and he began to sell Velcro in the United States in the 1950s. Over the years, Velcro has become a popular feature on shoes, coats, and other products.

Recognize, Recall, and Reflect

1. Why was George de Mestral so fascinated by the cockleburs he and his dog had come across while hunting in the mountains?

2. How does Velcro mimic a cocklebur?

3. What words were combined to make the word *Velcro*?

Investigate

In this activity, you will explore how hook-and-loop tape like Velcro works to fasten things together.

Materials

For each group of students:

- 1 in. strip of hook-and-loop tape

- Video microscope or dissecting microscope

- Different samples of fabric such as velvet, wool, and cotton

Safety Note: Do not handle wool if you are allergic to it. Wash your hands with soap and water immediately after completing this activity.

Create, Innovate, and Investigate

- Begin by observing the two different parts of the hook-and-loop tape. What do you notice about each part?

- Bend each piece of the tape and notice what happens to the piece with hooks.

- Try dragging the hooked tape across different fabrics such as velvet, a piece of wool, and a cotton sock. What do you observe?

- Watch what happens when you pull the hooks away from the different fabrics. Do the hooks bend? Do they break?

- Repeatedly scrape the hooks on a fabric to determine whether they lose the ability to fasten.

Questions for Reflection

1. What did you observe about hook-and-loop tape?

2. Is hook-and-loop tape equally effective in sticking to different types of fabric?

Apply and Analyze

How does Velcro work? The product is composed of two pieces of tape. One piece is covered with small hooks and the other piece is covered with small loops. As the two pieces of tape come together, the hooks connect with the loops, creating a bond. When the tape is pulled apart, the hooks flex and release the loops.

A new product designed by 3M Company uses a similar type of fastener with interlocking strips. But instead of having hooks or loops, both of the fastener's strips feature small, flexible pins that are shaped like mushrooms. The pins lock together when the two strips are pressed into each other. This new fastener is called Dual Lock fastener, and the makers of the product claim that it is stronger than Velcro. Find out more about these two fasteners by reviewing the sources listed below. Then answer the questions that follow.

Online Sources

- How does Velcro work?
  www.youtube.com/watch?v=mgcIivxODH0

- Velcro, observed under video microscope
  www.youtube.com/watch?v=62jhQNyPm3s

- 3M Scotch Extreme Dual Lock Fasteners review
  http://toolguyd.com/scotch-extreme-fasteners-review

1. What are the limitations of these types of fasteners when it comes to clothing?

2. How do you think the cost of Velcro compares to that of zippers or buttons?

3. Does Velcro last forever? What happens to it over time?

4. What is the advantage of a fastener with mushroom-shaped pins over a fastener with hooks and loops?

5. How could you design an experiment that would allow you to measure the strength of each of these fasteners?

Design Challenge

Engineering is the application of scientific understanding through creativity, imagination, problem solving, and the designing and building of new materials to address and solve problems in the real world. You will be asked to take the science you have learned in this case and design a process or product to address a real-world issue of your choosing.

Engineers use the engineering design process as steps to address a real-world problem (see Figure 22.2). You will now use this process as you come up with a new way to use hook-and-loop tape. In this case, you are asking the question (Step 1) of how you can design a new use for hook-and-loop tape. Drawing on your creativity, you will then brainstorm (Step 2) a new product that uses hook-and-loop tape to solve a problem. Afterward, you will create a plan (Step 3) for this new product. Next, you will create a sketch and/or model of your product (Step 4). Then, you will work with your classmates to think about how you would test (Step 5) and refine (Step 6) your product.

FIGURE 22.2

The Engineering Design Process

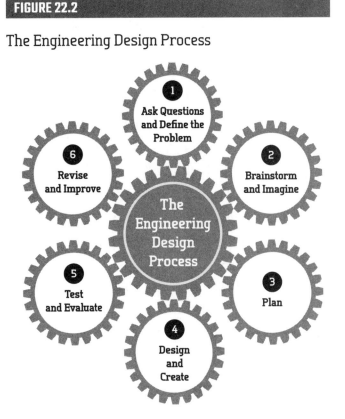

1. Ask Questions

Hook-and-loop tape serves many other purposes aside from being a fastener for clothing and shoes. For instance, the product can be used to hold down rugs, keep electrical cords coiled up, attach a remote control to a television, and much more. Based on your previous observations and research, consider a new problem that may be addressed or product that could be created using hook-and-loop tape. What are situations in which you would need to fasten objects together—and also easily be able to pull them apart? What problems could you solve using hook-and-loop tape?

2. Brainstorm and Imagine

Brainstorm a specific application for hook-and-loop tape that could help solve a problem. (For example, one idea is to use this tape to hold crib bumpers in place on the edge of a baby crib. Another idea is to use a biodegradable and edible type of hook-and-loop tape with food displays and tall, elaborate cakes.)

3. Create a Plan

Create a plan for your product. Consider: (1) What is the purpose of your product? (2) What are benefits to using your product? (3) What are the limitations of your product? Use the Product Planning Graphic Organizer to help you.

4. Design and Create

Consider the following questions and considerations for your product and its design.

- How would incorporating hook-and-loop tape into your design make the product better?

- How would you overcome any limitations or drawbacks caused by using hook-and-loop tape in your design?

- What technologies might need to be developed to create or manufacture this design?

- What are any constraints or drawbacks you can foresee with implementing this design?

- Would there be any safety concerns regarding your product?

Create a sketch of your product design. Make sure your design incorporates your previous research and exploration.

5. Test and Evaluate

Working with your classmates, come up with a way to test your design to see its effectiveness.

6. Revise and Improve

Give your plans to one of your classmates for review. Listen to his or her feedback on your design. What are some ways you can use the input to refine your design? Take some time to revise and make improvements.

Reflect

1. What technologies might need to be developed to create or manufacture this design?

2. What are any constraints or drawbacks you can foresee with implementing this design?

3. Would there be any environmental or human health concerns about this design?

Product Planning Graphic Organizer

Proposed Product Idea	
Pros (Benefits)	**Cons (Limitations)**

VELCRO

ENGINEERING MIMICS NATURE

A Case Study Using the Discovery Engineering Process

Lesson Overview

In this lesson, students explore Velcro and similar products that can be used to keep clothes and other objects fastened together. Velcro was an accidental discovery that has led to a number of new applications and products.

Lesson Objectives

By the end of this case study, students will be able to

- Describe the hook-and-loop system that makes up Velcro.

- Analyze how hook-and-loop tape functions to keep materials together and how it is able to repeatedly open and close.

- Design a new application for hook-and-loop tape.

The Case Study Approach

This lesson uses a case study approach. Explaining the purpose of case studies will encourage your students to relate to the material and engage with the problem. At the heart of each case study in this book is a true story, one that describes how someone in his or her everyday life or during a routine workday made an observation or did a simple experiment that led to a new insight or discovery. Case studies are designed to get students actively engaged in the process of problem solving. The narrative of the case supplies authentic details that place the student in the role of the inventor and provide scaffolds for critical thinking and deep reflection. A case is more than a paragraph to read or a story to analyze but rather a way of framing problems, synthesizing what is known, and thinking creatively about new applications and solutions. In this lesson, students consider how Velcro was discovered and work together to think about new applications for hook-and-loop tape to solve real-life problems.

Use of the Case

Due to the nature of these case studies, teachers may elect to use any section of each case for their instructional needs. The sections are sequenced in order (scaffolded) so students think more deeply about the science involved in the case and develop an understanding of engineering in the context of science.

Curriculum Connections

Related Next Generation Science Standards

PERFORMANCE EXPECTATIONS

- HS-ETS1-2. Design a solution to a complex real-world problem by breaking it down into smaller, more manageable problems that can be solved through engineering.

- HS-ETS1-3. Evaluate a solution to a complex real-world problem based on prioritized criteria and trade-offs that account for a range of constraints, including cost, safety, reliability, and aesthetics, as well as possible social, cultural, and environmental impacts.

SCIENCE AND ENGINEERING PRACTICES

- Analyzing and Interpreting Data
- Engaging in Argument From Evidence
- Constructing Explanations and Designing Solutions

Related National Academy of Engineering Grand Challenge

- Engineer the Tools of Scientific Discovery

Lesson Preparation

You will need to make copies of the entire student section for the class. Students will need internet access at various points in the lesson. Alternatively, you can project videos or print and distribute copies of online content for the class. For the Investigate section, you will need hook-and-loop tape. Sources for this material include fabric stores, hardware stores, and online shops. Look at the Teaching Organizer (Table 22.1, p. 372) for suggestions on how to organize the lesson.

Materials

For each group of students:

- 1 in. strip of hook-and-loop tape
- Video microscope or dissecting microscope
- Different samples of fabric such as velvet, wool, and cotton

Safety Note for Students: Do not handle wool if you are allergic to it. Wash your hands with soap and water immediately after completing this activity.

Time Needed

55 minutes

TABLE 22.1

Teaching Organizer

Section	Time Suggested	Materials Needed	Additional Considerations
The Case	5 minutes	Student packet	Could be read in class or as a homework assignment prior to class
Investigate	10 minutes	Student packet, 1-inch strip of hook-and-loop tape, video microscope or dissecting microscope, different samples of fabric such as velvet, wool, and cotton	Recommended as a small-group investigation
Apply and Analyze	10 minutes	Student packet, internet access	Whole-class activity or done individually
Design Challenge	30 minutes	Student packet	Small-group activity

Teacher Background Information

There are a number of resources and videos about hook-and-loop tape available on the internet. You may want to observe the behavior of hook-and-loop tape on sites such as YouTube prior to using the case.

Vocabulary

- biomimicry
- cocklebur
- fastener

Teacher Answer Key

Recognize, Recall, and Reflect

1. **Why was George de Mestral so fascinated by the cockleburs he and his dog had come across while hunting in the mountains?**

   He was intrigued by how effectively these seed pods clung to his pants.

2. **How does Velcro mimic a cocklebur ?**

   Part of the fastener has hooks that are modeled after the hooks on a cocklebur.

3. **What words were combined to make the word *Velcro*?**

   Velvet and crochet

Questions for Reflection

1. **What did you observe about hook-and-loop tape?**

   Answers will vary.

2. **Is hook-and-loop tape equally effective in sticking to different types of fabric?**

   No. Students should note that the tape generally sticks best to fabrics such as wool. It sticks less effectively to fabrics like cotton.

Apply and Analyze

1. **What are the limitations of these types of fasteners when it comes to clothing?**

   Answers will vary but may include that the small hooks on hook-and-loop tape can scratch or irritate skin. Also, hook-and-loop tape doesn't close two materials as tightly as other fasteners like buttons. Furthermore, the pressure that is needed to connect the hooks and loops of the fastener sometimes results in wrinkles.

2. **How do you think the cost of using Velcro compares to that of zippers or buttons?**

All three fastener systems can be made of synthetic materials and the cost is similar. Natural buttons or metal zippers could result in a costlier system.

3. **Does Velcro last forever? What happens to it over time?**

   If used over and over again, the hooked side of hook-and-loop tape can become entangled with fibers, hair, and other debris, making it difficult to maintain a tight connection with the looped side of the fastener.

4. **What is the advantage of a fastener with mushroom-shaped pins over a fastener with hooks and loops?**

   Students should note that the pins will not catch other materials (like a hair or fiber) as easily as a hook-and-loop system would.

5. **How could you design an experiment that would allow you to measure the strength of each of these fasteners?**

   Answers will vary, but students might suggest adding weights to pieces of material or fabric held together by the different fasteners to see how quickly each fastener comes undone under the stress of the weight.

Reflect

1. **What technologies might need to be developed to create or manufacture this design?**

   Answers may vary depending on the students' designs.

2. **What are any constraints or drawbacks you can foresee with implementing this design?**

   Answers may vary depending on the students' designs.

3. **Would there be any environmental or human health concerns?**

   Answers may vary depending on the students' designs.

Assessment

The Design Challenge can be assessed using the rubric in the appendix (p. 377).

Extensions

This lesson can be followed with lessons about the elasticity and adhesive properties of different types of glues. You can cover this topic with other cases in this book as well, including the case study in Chapter 10, "A Sticky Discovery: The Invention of Post-It Notes" (p. 153) and the case study in Chapter 17, "Super Glue: Accidentally Discovered Twice" (p. 269).

Resources and References

3M Industrial Assembly and Design. 2010. "3M™ Dual Lock™ Reclosable Fasteners: Bowling ball test." YouTube video. *www.youtube.com/watch?v=heavjPsXF1w.*

Anderson, T. 2002. "Velcro, observed under video microscope." YouTube video. *www.youtube.com/watch?v=62jhQNyPm3s.*

Design Squad Global. 2014. "How does Velcro work?" YouTube video. *www.youtube.com/watch?v=mgcIivxODH0.*

ToolGuyd. 2015. 3M Scotch Extreme Dual Lock Fasteners review. *http://toolguyd.com/scotch-extreme-fasteners-review.*

Appendix

Student Assessment

You may use the rubric below (or adapt it to your specific needs) when assessing the students' Design Challenge products in the case studies.

Rubric for Assessing the Design Challenges (36 Points Possible)

Product Engineering Construct	Not Proficient 1–2 Points	Proficient 3–4 Points	Strongly Proficient 5–6 Points
Ask Questions and Define the Problem	Student questions are limited and have little to no relevance to the case study; questions show a misunderstanding of the material.	Some student questions are meaningful and connected to the case study, but others may not be relevant or show limited understanding of the material.	Most student questions spark meaningful connections with the case study, prior learning, and personal experiences; questions show a deep understanding of the material and generate opportunities to transfer case study knowledge and skills to other situations.
Brainstorm and Imagine	Student product ideas are not relevant or they show a misunderstanding of the science behind the innovation featured in the unit.	Student product ideas are mostly relevant and show creativity; they incorporate some pertinent information from the unit; and they show a proficient understanding of the science behind the innovation featured in the unit.	Student product ideas are highly relevant and creative; they transfer information from the unit to other situations and contexts; and they show a deep understanding of the science behind the innovation featured in the unit.

Continued

Rubric for Assessing the Design Challenges (*continued*)

Product Engineering Construct	Not Proficient 1–2 Points	Proficient 3–4 Points	Strongly Proficient 5–6 Points
Create a Plan	Student plan is incomplete, disorganized, and/or irrelevant; the plan does not explain the purpose, benefits, and limitations of the product; it shows a lack of critical thinking.	Student plan is fairly thorough, organized, and relevant; it explains the purpose, benefits, and limitations of the product to a degree.	Student plan is very thorough, organized, and relevant; it explains in detail the purpose, benefits, and limitations of the product; it shows critical thinking skills.
Design and Create	Student design is poorly executed or incomplete; it does not account for relevant limitations or drawbacks; it does not adequately address the real-life problem the students sought to solve.	Student design is fairly well executed; it attempts to account for some relevant limitations and drawbacks; the design adequately addresses the real-life problem the students sought to solve; the design shows analysis of information and data at an appropriate level.	Student design is very well executed; it attempts to account for most relevant limitations and drawbacks; it is highly efficient in addressing the real-life problem the students sought to solve; the design process shows thorough analysis of information and data.
Test and Evaluate	Student evaluation/testing is limited and does not reveal the product's effectiveness.	Student evaluation/testing shows analysis of the design and its effectiveness; it helps students recognize flaws in the design and generate some ideas for improvement.	Student evaluation/testing shows critical analysis of the design and successfully reveals its effectiveness; the evaluation gives students a clear understanding of the product's drawbacks and limitations as well as ideas for improvement.
Revise and Improve	Students do not listen to feedback from others or do not use relevant suggestions to improve their designs.	Students listen to suggestions from their classmates and are able to make some improvements to their designs based on relevant feedback.	Students listen to feedback from their classmates and are able to utilize relevant suggestions very effectively in order to make improvements to their designs.

Image Credits

All art by NSTA Press unless otherwise noted.

Chapter 1

Figure 1.1: Petermaerki, Wikimedia Commons, CC BY-SA 3.0, *https://commons.wikimedia. org/wiki/File:Nitinol_bueroklammer_verbogen.jpg*

Figure 1.2: Kihopczmaluoch, Wikimedia Commons, CC BY-SA 4.0, *https://commons. wikimedia.org/wiki/File:HK_while_damaged_cable_USB_Apple_Type-A_plugs_iPhone_iPad_ accessories_hardware_connector_%E5%A1%91%E8%86%A0%E7%A1%AC%E5%8C%96_ Plastic_hardening_Power_charger_data_line_%E9%87%91%E5%B1%AC%E7%96%B2%E5 %8B%9E_Metal_fatigue_May_2017_IX1_02.jpg*

Chapter 2

Figure 2.1 (left): Didier Descouens, Wikimedia Commons, CC-BY-SA-4.0, *https:// commons.wikimedia.org/wiki/File:Morpho_rhetenor_rhetenor_MHNT.jpg*

Figure 2.1 (right): Didier Descouens, Wikimedia Commons, CC-BY-SA-3.0, *https:// commons.wikimedia.org/wiki/File:Danaus_plexippus_MHNT_femelle_dos.jpg*

Figure 2.2: Juliano Costa, Wikimedia Commons, CC-BY-SA-3.0, *https://commons. wikimedia.org/wiki/File:Pollia.jpg*

Figure 2.3: Richard Bowdler Sharpe, Wikimedia Commons, Public domain. *https:// commons.wikimedia.org/wiki/File:Parotia_lawesii_by_Bowdler_Sharpe.jpg*

Chapter 3

Figure 3.1: LoggaWiggler, Pixabay, CC0 1.0, *https://pixabay.com/en/dyeing-color-factory- wool-fabric-15038*

Chapter 4

Figure 4.1: Figure: Pascal Deynat/Odontobase, Wikimedia Commons, CC BY AS 3.0, *https://commons.wikimedia.org/wiki/File:Denticules_cutan%C3%A9s_du_requin_citron_ Negaprion_brevirostris_vus_au_microscope_%C3%A9lectronique_%C3%A0_balayage.jpg*

Chapter 5

Figure 5.1: Shutterstock

Figure 5.2: Shutterstock

Chapter 6

Figure 6.1: Authors

Figure 6.2: From Porter, M. et al. 2015. Why the seahorse tail is square. *Science* 349 (6243): aaa6683. Reprinted with permission from AAAS.

Figure 6.3: From Porter, M. et al. 2015. Why the seahorse tail is square. *Science* 349 (6243): aaa6683. Reprinted with permission from AAAS.

Chapter 7

Corn flakes: Suzette-www.suzette.nu, Wikimedia Commons, CC BY 2.0, *https://commons.wikimedia.org/wiki/File:Corn_flakes_(3862325873).jpg*

Chapter 8

Match: Heidas, Wikimedia Commons, CC BY-SA 3.0, *https://commons.wikimedia.org/wiki/File:Streichholz.JPG*

Figure 8.1: Shutterstock

Chapter 9

Figure 9.1: MichaelGaida, Pixabay, CC0 1.0, *https://pixabay.com/en/phone-old-telephone-handset-2863663*

Chapter 10

Sticky note: Disk Depot, Wikimedia Commons, CC BY-SA 3.0, *https://commons.wikimedia.org/wiki/File:Post-it-note-transparent.png*

Figure 10.1: Authors

Chapter 11

Girl with putty: University of the Fraser Valley, Wikimedia Commons, CC BY 2.0, *https://commons.wikimedia.org/wiki/File:UFV_Science_Rocks_(14534714507).jpg*

Chapter 12

Saccharin on spoon: Shutterstock

Figure 12.1: Gert-Wolfhard von Rymon Lipinski, Public domain. *https://commons.wikimedia.org/wiki/File:Remsen-Fahlberg_synthesis_of_saccharin.png*

Chapter 13

Broken glass piece: Sardaka, Wikimedia Commons, CC-BY-3.0, *https://commons.wikimedia. org/wiki/File:(1)Myuna_Creek_broken_bottle.jpg*

Figure 13.1: Shutterstock

Figure 13.2: Shutterstock

Figure 13.3: Georg Slickers, Wikimedia Commons, CC BY-SA 2.5, *https://commons.wikimedia. org/wiki/File:Safety_glass_vandalised_20050526_062_part.jpg*

Chapter 14

Figure 14.1: Anwar Huq (University of Maryland Biotechnology Institute, Baltimore, Maryland, United States), Wikimedia Commons, CC BY 2.5, *https://commons.wikimedia.org/ wiki/File:Washing_Utensils_And_Vegetables.png*

Figure 14.2: Shutterstock

Figure 14.3: Shutterstock

Figure 14.4: Western Nanofabrication Facility, University of Western Ontario, used with permission, *http://nanofabrication.tumblr.com/post/91041117461/nylon-stockings-under-the-sem*

Chapter 15

Slinky: Roger McLassus, Wikimedia Commons, CC BY-SA 3.0, *https://commons.wikimedia. org/wiki/File:2006-02-04_Metal_spiral.jpg*

Chapter 16

Figure 16.1 (left): Authors

Figure 16.1 (right): Authors

Chapter 17

Superglue: Omegatron, Wikimedia Commons, CC BY-SA 3.0, *https://commons.wikimedia.org/ wiki/File:Super_glue.jpg*

Chapter 18

Figure 18.1: Shutterstock

Chapter 19

Figure 19.1: Tyne & Wear Archives & Museums, Wikimedia Commons, Public domain. *https://commons.wikimedia.org/wiki/File:Vaseline_-_TWCMS-G12142_(16692709641).jpg*

Chapter 20

Shoe: Ayo Ogunseinde armedshutter, Wikimedia Commons, CC0 1.0, *https://commons. wikimedia.org/wiki/File:Houston_sneakers_(Unsplash).jpg*

Chapter 21

Figure 21.1 (left): D-Kuru, Wikimedia Commons, CC BY-SA 3.0 AT, *https://commons. wikimedia.org/wiki/File:Crookes_tube-not_in_use-lateral_view_prPNr%C2%B002.jpg*

Figure 21.1 (right): D-Kuru, Wikimedia Commons, CC BY-SA 3.0 AT, *https://commons. wikimedia.org/wiki/File:Crookes_tube-in_use-lateral_view-2_prPNr%C2%B009.jpg*

Figure 21.2: Wilhelm Röntgen, Wikimedia Commons, Public domain. *https://commons. wikimedia.org/wiki/File:Room_where_R%C3%B6ntgen_found_x-rays.jpg*

Figure 21.3: Wikimedia Commons, Public domain. *https://commons.wikimedia.org/ wiki/File:First_medical_X-ray_by_Wilhelm_R%C3%B6ntgen_of_his_wife_Anna_Bertha_ Ludwig%27s_hand_-_18951222.jpg*

Figure 21.4: Authors

Chapter 22

Figure 22.1: Cocklebur, Stan Shebs, Wikimedia Commons, CC BY-SA 3.0, *https://commons. wikimedia.org/wiki/Xanthium_strumarium#/media/File:Xanthium_strumarium_5.jpg*

Velcro: Ryj, Wikimedia Commons, CC BY-SA 3.0, https://commons.wikimedia.org/wiki/ Category:Velcro#/media/File:Klettverschluss.jpg

Index

Note: Page references in **boldface** indicate information contained in figures or tables.

A

acid, 38
Acinetobacter baumannii, **61**
adhesion, 166
adhesive, 85, 166, 283
Advance Health Informatics (Grand Challenge)
    memory wire, 20
    Vaseline, 319
alloy, 22, 266
ammonia, 195
Analyzing and Interpreting Data
    artificial dye, 52
    corn flakes, 119
    gecko feet adhesives, 83
    matches, 133
    memory wire, 20
    plastic, 147
    Post-it notes, invention of, 162
    saccharin sweetener, 193
    safety glass, 209
    saris, and cholera, 229
    seahorse tails, 99
    Silly Putty, discovery of, 179
    Slinky, the, 247
    stainless steel, 263
    Super Glue, 281
    Teflon, 303
    Vaseline, 318
    Velcro, 373
    vulcanized rubber, 334
    x-rays, discovery of, 356
aneurysm, 283
annealed, 212

anode, 358
antimicrobial, 195
articulated, 102
artificial dye (case study)
    apply and analyze, 46–47
    assessment, 56
    case, 43–44, **44**
    curriculum connections, 52
    design challenge, **47**, 47–49
    extensions, 56
    introduction, 43
    investigate, 45–46, **46**
    lesson integration, 52
    lesson objectives, 43, 51
    lesson overview, 51
    lesson preparation, 53–54, **54**
    related National Academy of Engineering Grand Challenge, 52
    related *Next Generation Science Standards*, 52
    resources and references, 56
    teacher answer key, 55–56
    teacher notes, 51–55
artificial sweetener, 195
Asking Questions and Defining Problems
    safety glass, 209
    shark skin and bacteria, 68
    Vaseline, 318
    x-rays, discovery of, 356
assessment rubric, 377–378
atomic forces, 85
atomic number, 358

atoms, 22, 306, 358

B
bacteria, 230
Baekeland, Leo, 139–140
Bakelite, 140
base, 38
Benedictus, Edouard, 200
benzoic sulfimide, 195
biology case studies, **10**
biomimetics, 102, 373
blue morpho butterfly, **26, 29**
bond, 306
bonding, 85
Brearley, Harry, 254
Brennan, Anthony, 58
Buehler, William, 12, 14

C
Cambrian explosion, 29
casein, 306
case study approach, 2–4
cathode, 358
Cause and Effect (crosscutting concept)
    matches, 133
    safety glass, 210
    seahorse tails, 99
    structural color and iridescence, 37
    Vaseline, 319
    vulcanized rubber, 335
    x-rays, discovery of, 356
cellulose, 38
cellulose nitrate, 212
cereal grain, 120
chemistry case studies, **10**
Chesebrough, Robert Augustus, 310
cholera, 217, 218–219, **222,** 222–223, 230. *See also* saris, and cholera (case study)
chronic obstructive pulmonary disease (COPD), 250
cocklebur, **363,** 373
cohesion, 166
coil spring, 249
Colwell, Rita, 218–219
compound, 166

Constructing Explanations and Designing Solutions
    artificial dye, 52
    corn flakes, 119
    gecko feet adhesives, 83
    matches, 133
    memory wire, 20
    plastic, 147
    Post-it notes, invention of, 162
    saccharin sweetener, 193
    safety glass, 209
    saris, and cholera, 229
    seahorse tails, 99
    Silly Putty, discovery of, 179
    Slinky, the, 247
    stainless steel, 263
    structural color and iridescence, 36
    Super Glue, 281
    Teflon, 303
    Vaseline, 319
    Velcro, 373
    vulcanized rubber, 334
    x-rays, discovery of, 356
converted, 306
Coover, Harry, 270
copepod, 230
corn flakes (case study)
    apply and analyze, 114
    assessment, 122
    case, 106–107
    Create Your Own Cereal, 111–113
    curriculum connections, 119
    design challenge, 114–116
    design improvement chart, 110
    extensions, 122
    introduction, 105–106
    investigate, 107–109, **108–109**
    lesson objectives, 106, 118
    lesson overview, 118
    lesson preparation, 120
    related National Academy of Engineering Grand Challenge, 119
    related *Next Generation Science Standards*, 119
    resources and references, 122–123
    teacher answer key, 121–122

teacher notes, 118–123
corrosive, 306
covalent, 85
Crosby, Alfred, 74
cross section, 102
crystal, 22
current, 358
cyanoacrylate, 283

D
de Mestral, George, 1, 364
desiccant, 266
design challenges assessment rubric, 377–378
Developing and Using Models
  seahorse tails, 99
    shark skin and bacteria, 68
    x-rays, discovery of, 356
diffract, 358
diffraction pattern, 358
discovery engineering
  about, 1–2, 8
dissociate, 266

E
Earth/space science case studies, **10**
elasticity, 166
electrode, 358
electrolyte, 266
electromagnetic radiation, 358
electron, 358
electronegativity, 195
electrostatic attraction, 166
energy, 249
Energy and Matter (crosscutting concept)
  corn flakes, 119
  matches, 133
  Slinky, the, 247
energy level, 358
Engaging in Argument From Evidence
  corn flakes, 119
  gecko feet adhesives, 83
  memory wire, 20
  Post-it notes, invention of, 162
  saccharin sweetener, 193
  safety glass, 209

saris, and cholera, 229
seahorse tails, 99
Silly Putty, discovery of, 179
Slinky, the, 247
stainless steel, 263
structural color and iridescence, 36
Super Glue, 281
Vaseline, 319
Velcro, 373
vulcanized rubber, 334
x-rays, discovery of, 356
Engineer Better Medicines (Grand Challenge)
  memory wire, 20
  Vaseline, 319
engineering controls, 9
engineering design process, 6, **7, 79**
Engineer the Tools of Scientific Discovery (Grand Challenge)
  artificial dye, 52
  corn flakes, 119
  gecko feet adhesives, 84
  matches, 133
  memory wire, 20
  plastic, 147
  Post-it notes, invention of, 162
  saccharin sweetener, 194
  safety glass, 210
  Silly Putty, discovery of, 179
  Slinky, the, 247
  stainless steel, 264
  structural color and iridescence, 37
  Super Glue, 281
  Teflon, 303
  Vaseline, 319
  Velcro, 373
  vulcanized rubber, 335
erode, 306
evaporative cooling, 306

F
Fahlberg, Constantin, 186
Fallgatter, Ruth, 168
fastener, 373
flavin, 38
fluorescent/fluoresce, 358

fluorocarbon, 306
Ford Model A, **201**
formaldehyde, 149
Franklin, Rosalind, 349, 360
Fry, Arthur, 154

G
gecko feet adhesives (case study)
    Adhesive Testing Chart, 77
    apply and analyze, 78
    assessment, 87
    case, 74–75
    curriculum connections, 83–84
    design challenge, 78–80
    extensions, 87
    introduction, 73
    investigate, 75–78
    lesson objectives, 73, 82
    lesson overview, 82
    lesson preparation, 84–85, **85**
    related National Academy of
        Engineering Grand Challenge,
        84
    related *Next Generation Science
        Standards*, 83
    resources and references, 88
    teacher answer key, 86–87
Goodyear, Charles, 326–327
Grand Challenges for Engineering for
    the 21st century
    about, 7–8

H
helical spring, 249
heterogeneous, 181
Hodgson, Peter, 168
Hoskissen, Kirsten, 114
hydrophobic, 306, 321

I
indicator, 54
infections, common healthcare-
    associated, **61**
ionic, 85
iridescence/iridescent, 38
Irschick, Duncan, 74
isomer, 195

J
James, Richard, 236
Jones, Samuel, 126
Joyner, Fred, 270

K
Kellogg family, 106–107
kinetic energy, 249
Klebsiella pneumoniae, **61**

L
lab safety, 9
laminated, 212
latex, 337
Lawes's parotia bird, **30**
Lundstrom, Johan Edvard, 126

M
maize, 120
malt, 120
matches (case study)
    apply and analyze, 128
    case, 126, **126**
    design challenges, 128–130
    introduction, 125
    investigate, **127,** 127–128
    lesson objectives, 125, 132
    lesson overview, 132
    related National Academy of
        Engineering Grand Challenge,
        133
    related *Next Generation Science
        Standards*, 133
    teacher notes, 131–137
memory wire (case study)
    apply and analyze, 14–15
    assessment, 24
    case, 11
    crosscutting concept, 20
    curriculum connections, 20
    design challenge, 15–17, **16**
    extensions, 24
    introduction, 11
    investigate, 13–14
    lesson integration, 20
    lesson objectives, 11, 19
    lesson overview, 19
    lesson preparation, 21–22, **22**

related National Academy of Engineering Grand Challenges, 20

related *Next Generation Science Standards*, 20

resources and references, 24

science and engineering practices, 20

teacher answer key, 23–24

teacher notes, 19–24

metal, 22

metallic elements, 195

metallurgist, 22

metals, properties of, 22

milled corn, 120

mixture, 181

molecule, 166

monomer, 306

Muzzey, David S., 12

Mycobacterium abscessus, **61**

N

nano-, 38

nanoparticles, 7

Next Generation Science Standards (NGSS), 4–7, **5**

Nitinol wire, 22

nonmetallic elements, 195

non-Newtonian fluid, 181

non-nutritive, 195

nonstick coating, 306

Norovirus, **61**

O

octet rule, 195

opaque/opacity, 358

orbital, 358

oxidation, 195, 266

P

Panati, Charles, 310

Pasch, Gustaf Erik, 126

Patterns (crosscutting concept)

matches, 133

Vaseline, 319

Perkins, William, 43, 44

personal protective equipment, 9

petroleum, 321

petroleum jelly, 309. *See also* Vaseline (case study)

pH, 38

phase change, 22

phenol, 149

phosphorus chloride, 195

physics case studies, **10**

phytoplankton, 230

pigment, 38

plankton, 230

Planning and Carrying Out Investigations

x-rays, discovery of, 356

plastic, 306

plastic (case study)

apply and analyze, 142

assessment, 151

case, 139–140

curriculum connections, 147

design challenge, 142–144, **143**

extensions, 151

introduction, 139

investigate, 141–142

lesson integration, 147

lesson objectives, 139, 146

lesson overview, 146

lesson preparation, 148–149, **149**

related National Academy of Engineering Grand Challenges, 147

related *Next Generation Science Standards*, 147

resources and references, 151

teacher answer key, 150–151

teacher notes, 146–151

plastic polymer, 306

Plunkett, Roy, 288

Pollia condensata, **26**, 27

polymer, 149, 283, 306, 337

polyvinyl butyral (PVB), 212

Post-it notes, invention of (case study)

apply and analyze, 155–157

assessment, 165

case, 153–154

curriculum connections, 162

design challenge, 157–159

extensions, 166

introduction, 153

investigate, 154–155, **155**
lesson integration, 162
lesson objectives, 153, 161
lesson overview, 161
lesson preparation, **163,** 163–164
product planner graphic
    organizer, 160
related National Academy of
    Engineering Grand Challenge,
    162
related *Next Generation Science
    Standards,* 162
resources and references, 166
teacher answer key, 164–165
teacher notes, 161–166
potential energy, 249
prehensile, 102
Provide Access to Clean Water (Grand
    Challenge)
    saris, and cholera, 229
Pseudomonas aeruginosa, **61**

R

radiograph, 358
radiological density, 358
redox, 195
reduction, 195, 266
refrigerant, 306
Remsen, Ira, 186
resin, 149
Restore and Improve Urban
    Infrastructure (Grand Challenge)
    safety glass, 210
    vulcanized rubber, 335
reverse engineering, 1–2
roller, 120
Röntgen, Wilhelm Conrad, 342–344,
    343, 344

S

saccharin sweetener (case study)
    apply and analyze, **189,** 189–190
    assessment, 197
    case, 186
    curriculum connections, 193–194
    design challenge, 190–191
    extensions, 197
    introduction, 185–186

investigate, 187–189
lesson integration, 193
lesson objectives, 186, 192
lesson overview, 192
lesson preparation, 194–195, **195**
related National Academy of
    Engineering Grand Challenge,
    194
related *Next Generation Science
    Standards,* 193–194
resources and references, 197–198
teacher answer key, 196–197
teacher notes, 192–198
safety glass (case study)
    apply and analyze, 202–204, **203,
        204**
    assessment, 215
    case, 200–201, **201**
    curriculum connections, 209–210
    design challenge, 204–206
    extenstions, 215
    introduction, 199
    investigate, 201–202
    lesson integration, 209
    lesson objectives, 199–200, 208
    lesson overview, 208
    lesson preparation, 210–212, **211**
    related National Academy of
        Engineering Grand Challenges,
        210
    related *Next Generation Science
        Standards,* 209–210
    resources and references, 216
    teacher answer key, 213–215
    teacher notes, 208–216
safety practices, 9
saris, and cholera (case study)
    apply and analyze, **222,** 222–223
    assessment, 232
    case, 218–219
    characteristics of fabrics chart, 221
    curriculum connections, 228–229
    design challenge, 223–225
    extensions, 232
    introduction, 217, **217**
    investigate, 219–222, **220**
    lesson integration, 228
    lesson objectives, 218, 227

lesson overview, 227

lesson preparation, 229–230, **230**

related National Academy of Engineering Grand Challenge, 229

related *Next Generation Science Standards*, 228–229

resources and references, 232–233

teacher answer key, 231–232

teacher notes, 227–233

Scale, Proportion, and Quantity (crosscutting concept)

structural color and iridescence, 37

x-rays, discovery of, 356

scientific method, 5–6, **6**

seahorse tails (case study)

apply and analyze, 93–94

assessment, 103

case, 90–92, **91**

curriculum connections, 99

design challenge, **94,** 94–96

extensions, 104

introduction, 89

investigate, 92–93

lesson objectives, 90, 98

lesson overview, 98

lesson preparation, 100–102, **101**

related *Next Generation Science Standards*, 99

resources and references, 104

teacher answer key, 102–103

setae, 85

sharklet material, 58, **59**

shark skin and bacteria (case study)

apply and analyze, 60–62, **61**

assessment, 71–72

Business Plan Form, 65

case, 58–59, **59**

curriculum connections, 68

design challenge, **62,** 62–64

extensions, 72

introduction, 57

investigate, 59–60, **60**

lesson integration, 68

lesson objectives, 57, 67

lesson overview, 67

lesson preparation, 68–70, **69**

New Product Evaluation Form, 66

related *Next Generation Science Standards*, 68

resources and references, 72

teacher answer key, 70–71

teacher notes, 67–72

shellac, 139–140

Silly Putty, discovery of (case study)

apply and analyze, 171–172

assessment, 183

case, 168

curriculum connections, 178–179

design challenge, 172–175

extensions, 183

introduction, 167

investigate, 169

lesson integration, 178

lesson objectives, 167, 177

lesson overview, 177

lesson preparation, 179–181, **181**

make your own putty, 169–171

related National Academy of Engineering Grand Challenge, 179

related *Next Generation Science Standards*, 178–179

resources and references, 183–184

teacher answer key, 182–183

teacher notes, 177–184

Silver, Spencer, 153–154

Slinky, the (case study)

apply and analyze, 242

assessment, 251

case, 236

crosscutting concepts, 247

curriculum connections, 247–248

design challenge, 242–244

extensions, 251

introduction, 235

investigate, **237,** 237–242, **240–241**

lesson integration, 247

lesson objectives, 236, 246

lesson overview, 246

lesson preparation, 248–249, **249**

related National Academy of Engineering Grand Challenge, 248

related *Next Generation Science Standards*, 247
resources and references, 251–252
teacher answer key, 249–251
teacher notes, 246–252
solution, 38
square prism, 102
Stability and Change (crosscutting concept)
Teflon, 303
stabilizer, 337
stainless steel (case study)
apply and analyze, 258
assessment, 268
case, 254–255
curriculum connections, 263–264
design challenge, 258–260
extensions, 268
introduction, 253
investigate, **255,** 255–258
lesson integration, 263
lesson objectives, 253–254, 262
lesson overview, 262
lesson preparation, **264,** 264–266, **266**
related National Academy of Engineering Grand Challenge, 264
related *Next Generation Science Standards*, 263
resources and references, 268
teacher answer key, 266–268
teacher notes, 262–268
Staphylococcus aureus, **61**
staple, 120
states of matter, 181
structural color and iridescence (case study)
apply and analyze, 29–30, **30**
assessment, 41
case, **26,** 26–27
crosscutting concepts, 37
curriculum connections, 36–37
design challenge, **31,** 31–33
extensions, 41
introduction, 25, **26**
investigate, 27–29
lesson integration, 36

lesson objectives, 25, 35
lesson overview, 35
lesson preparation, **37,** 37–38
related National Academy of Engineering Grand Challenge, 37
related *Next Generation Science Standards*, 36
resources and references, 41
science and engineering practices, 36
teacher answer key, 39–40
teacher notes, 35–41
Structure and Function (crosscutting concept)
artificial dye, 52
corn flakes, 119
gecko feet adhesives, 83
memory wire, 20
plastic, 147
saccharin sweetener, 194
safety glass, 210
seahorse tails, 99
Silly Putty, discovery of, 179
Slinky, the, 247
stainless steel, 263
structural color and iridescence, 37
Super Glue, 281
Teflon, 303
Vaseline, 319
x-rays, discovery of, 356
sucrose, 195
sulfobenzoic acid, 195
sulfobenzoic compounds, 195
Super Glue (case study)
apply and analyze, 276
assessment, 285
case, 270
curriculum connections, 281
design challenge, 276–278
extensions, 285
introduction, 269, **269**
investigate, 270–276, **272, 274–275**
lesson integration, 281
lesson objectives, 269–270, 280
lesson overview, 280
lesson preparation, 282–283

related National Academy of Engineering Grand Challenge, 281

related *Next Generation Science Standards*, 281

resources and references, 285

teacher answer key, 283–285

teacher notes, 280–285

suspension, 181

synthetic, 149

Systems and System Models (crosscutting concept)
   Slinky, the, 247

T

tack, 85

Teflon (case study)
   apply and analyze, 297
   assessment, 307
   case, 288
   curriculum connections, 302–303
   design challenge, 297–299
   dive deeper, 289–292, **290–291, 293**
   extensions, 307
   introduction, **287,** 287–288
   investigate, 294–295, **296**
   lesson integration, 302
   lesson objectives, 288, 301
   lesson overview, 301
   lesson preparation, 303–306, **305**
   related *Next Generation Science Standards*, 302–303
   resources and references, 308
   teacher answer key, 306–307
   teacher notes, 301–308

tempered, 212

thermoset, 149

thin film, 38

Thomson, J.J., 344

V

valence, 195

van der Waals forces, 85, 166

Vaseline (case study)
   apply and analyze, 313
   assessment, 323
   case, **310,** 310–311
   crosscutting concepts, 319

curriculum connections, 318–319

design challenge, 313–315

extensions, 323

introduction, 309

investigate, 311–313

lesson integration, 318

lesson objectives, 309, 317

lesson overview, 317

lesson preparation, 319–321, **320**

related National Academy of Engineering Grand Challenge, 319

related *Next Generation Science Standards*, 318–319

resources and references, 323

teacher answer key, 321–323

teacher notes, 317–323

vegetarian, 120

Velcro, 1

Velcro (case study)
   apply and analyze, 366
   assessment, 374
   case, **364,** 364–365
   curriculum connections, 371
   design challenge, 366–368, **367**
   extensions, 374–375
   introduction, 363, **363**
   investigate, 365
   lesson objectives, 364, 370
   lesson overview, 370
   lesson preparation, 371–373, **372**
   related National Academy of Engineering Grand Challenge, 371
   related *Next Generation Science Standards*, 371
   resources and references, 375
   teacher answer key, 373–374
   teacher notes, 370–375

Vignolini, Silvia, 26–27

viscosity, 181

von Laue, Max, 348

vulcanized rubber (case study)
   apply and analyze, 328–329
   assessment, 338
   case, 326–327
   curriculum connections, 334–335
   design challenge, 329–331

extensions, 339
introduction, 325
investigate, 327–328
lesson integration, 334
lesson objectives, 325, 333
lesson overview, 333
lesson preparation, 335–337, **336**
related National Academy of
   Engineering Grand Challenges,
   335
related *Next Generation Science
   Standards*, 334–335
resources and references, 339
teacher answer key, 337–338
teacher notes, 333–339

W
Walker, John, 126
water-soluble, 195
wavelength, 38
Wright, James, 168

X
x-rays, discovery of (case study)
   apply and analyze, 348–349

assessment, 361
case, **342,** 342–344, **343**
crosscutting concepts, 356
curriculum connections, 355–356
design challenge, 349–351, **353**
dive deeper, 344–345
extensions, 361
introduction, 341
investigate, 345–348, **347**
lesson integration, 355
lesson objectives, 341, 354
lesson overview, 354
lesson preparation, 356–358, **357**
related National Academy of
   Engineering Grand Challenges,
   356
related *Next Generation Science
   Standards*, 355–356
resources and references, 361
teacher answer key, 358–360

Z
zooplankton, 230